PRINCIPLES AND PRACTICE OF ENGINEERING
CIVIL ENGINEERING
SAMPLE QUESTIONS & SOLUTIONS

Published by the
National Council of Examiners for Engineering and Surveying®
280 Seneca Creek Road, Clemson, SC 29631 800-250-3196 www.ncees.org

ISBN 1-932613-10-2

Printed in the United States of America

TABLE OF CONTENTS

INTRODUCTION

The National Council of Examiners for Engineering and Surveying (NCEES) has prepared this handbook to assist candidates who are preparing for the Principles and Practice of Engineering (PE) examination in civil engineering. The NCEES is an organization established to assist and support the licensing boards that exist in all states and U.S. territories. One of the functions of the NCEES is to develop examinations that are taken by candidates for licensure as professional engineers. The NCEES then provides the licensing boards with uniform examinations that are valid measures of minimum competency related to the practice of engineering.

To develop reliable and valid examinations, the NCEES employs procedures using the guidelines established in the Standards for Educational and Psychological Testing published by the American Psychological Association. These procedures are intended to maximize the fairness and quality of the examinations. To ensure that the procedures are followed, the NCEES uses experienced testing specialists possessing the necessary expertise to guide the development of examinations using current testing techniques.

Committees composed of professional engineers from throughout the nation prepare the examinations. These licensed engineers supply the content expertise that is essential in developing examinations. By using the expertise of licensed engineers with different backgrounds such as private consulting, government, industry, and education, the NCEES prepares examinations that are valid measures of minimum competency.

LICENSING REQUIREMENTS

ELIGIBILITY

The primary purpose of licensure is to protect the public by evaluating the qualifications of candidates seeking licensure. While examinations offer one means of measuring the competency levels of candidates, most licensing boards also screen candidates based on education and experience requirements. Because these requirements vary between boards, it would be wise to contact the appropriate board. Board addresses and telephone numbers may be obtained by visiting our Web site at www.ncees.org or calling (800) 250-3196.

APPLICATION PROCEDURES AND DEADLINES

Application procedures for the examination and instructional information are available from individual boards. Requirements and fees vary among the boards, and applicants are responsible for contacting their board office. Sufficient time must be allotted to complete the application process and assemble required data.

DESCRIPTION OF EXAMINATIONS

EXAMINATION SCHEDULE

The NCEES PE examination in civil engineering is offered to the boards in the spring and fall of each year. Dates of future administrations are as follows:

Year	Spring Dates	Fall Dates
2004	April 16	October 29
2005	April 15	October 28
2006	April 21	October 27

You should contact your board for specific locations of exam sites.

EXAMINATION CONTENT

The 8-hour PE examination in civil engineering is administered in two 4-hour sessions each containing 40 questions. Each question has four answer options. The morning session covers the breadth of civil engineering, and the afternoon session presents five alternate modules from which the examinee selects one. The five afternoon modules are environmental, geotechnical, structural, transportation, and water resources. The examination specifications presented in this book give details of the subjects covered on the examination.

Questions are either single independent questions or grouped in sets of up to three questions. A set of questions is preceded by information or data that is common to the questions that follow. The questions supply any further information specific to that question and define what is expected as a response to the question. The response required involves a calculation and/or conclusion that demonstrates competent engineering judgment.

This book presents half the number of questions that appear in an actual examination. By illustrating the general content of the topic areas, the questions should be helpful in preparing for the examination. Solutions are presented for all the questions. The solution presented may not be the only way to solve the question. The intent is to demonstrate the typical effort required to solve each question.

No representation is made or intended as to future examination questions, content, or subject matter.

EXAMINATION DEVELOPMENT

EXAMINATION VALIDITY

Testing standards require that the questions on a licensing examination be representative of the important tasks needed for competent practice in the profession as a licensed professional. The NCEES establishes the relationship between the examination questions and tasks by conducting an analysis of licensed practitioners that identifies the duties performed by the engineer. This information is used to develop an examination content outline that guides the development of job-related questions.

EXAMINATION SPECIFICATIONS

The examination content outline presented in this book specifies the subject areas that were identified for civil engineering and the percentage of questions devoted to each of them. The percentage of questions assigned to each of the topic areas reflects both the frequency and importance experienced in the practice of civil engineering.

EXAMINATION PREPARATION AND REVIEW

Examination development and review workshops are conducted frequently by standing committees of the NCEES. Additionally, question-writing workshops are held as required to supplement the bank of questions available. The content and format of the questions are reviewed by the committee members for compliance with the specifications and to ensure the quality and fairness of the examination. These licensed engineers are selected with the objective that they be representative of the profession in terms of geography, ethnic background, gender, and area of practice.

MINIMUM COMPETENCY

One of the most critical considerations in developing and administering examinations is the establishment of passing scores that reflect a standard of minimum competency. The concept of minimum competency is uppermost in the minds of the committee members as they assemble questions for the examination. Minimum competency, as measured by the examination component of the licensing process, is defined as the lowest level of knowledge at which a person can practice professional engineering in such a manner that will safeguard life, health, and property and promote the public welfare.

To accomplish the setting of fair passing scores that reflects the standard of minimum competency, the NCEES conducts passing score studies. At these studies, a representative panel of engineers familiar with the examinee population uses established procedures to set the passing score for the examination. Such procedures are widely recognized and accepted for occupational licensing purposes. The panel discusses the concept of minimum competence and develops a written standard of minimum competency that clearly articulates what skills and knowledge are required of licensed engineers. Following this, the panelists take the examination and then evaluate the difficulty level of each question in the context of the standard of minimum competency.

The NCEES does not use a fixed-percentage pass rate such as 70% or 75% because licensure is designed to ensure that practitioners possess enough knowledge to perform professional activities in a manner that protects the public welfare. The key issue is whether an individual candidate is competent to practice and not whether the candidate is better or worse than other candidates.

The passing score can vary from one administration of the examination to another to reflect differences in difficulty levels of the examinations. However, the passing score is always based on the standard of minimum competency. To avoid confusion that might arise from fluctuations in the passing score, scores are converted to a standard scale which adopts 70 as the passing score. This technique of converting to a standard scale is commonly employed by testing specialists. Some licensing jurisdictions may choose to report examination results on a pass/fail basis and not provide a numeric score.

SCORING PROCEDURES

The examination consists of 80 equally weighted multiple-choice questions. There is no penalty for marking incorrect responses; therefore candidates should answer each question on the examination. Only one response should be marked for each question. No credit is given where two or more responses are marked. The examination is compensatory—poor scores in some topics can be offset by superior performance elsewhere.

The legal authority for making licensure decisions rests with the individual licensing boards and not with the NCEES. Consequently, each board has the authority to determine the passing score for the examination. The NCEES provides each board with a recommended passing score based on the procedures described previously.

EXAMINATION PROCEDURES AND INSTRUCTIONS

EXAMINATION MATERIALS

Before the morning and afternoon sessions, proctors will distribute examination booklets containing an answer sheet. You should not open the examination booklet until you are instructed to do so by the proctor. Read the instructions and information given on the front and back covers and enter your name in the upper right corner of the front cover. Listen carefully to all the instructions the proctor reads. The proctor has final authority on the administration of the examination.

The answer sheets for the multiple-choice questions are machine scored. For proper scoring, the answer spaces should be blackened completely. Use only #2 pencils or mechanical pencils with HB lead. Marks in ink or felt-tip pens may not be scanned properly. If you decide to change an answer, you must erase the first answer completely. Incomplete erasures and stray marks may be read as intended answers. One side of the answer sheet is used to collect identification and biographical data that may be used to analyze the performance of the examination. The biographical data has no impact on the examination score. Proctors will guide you through the process of completing this portion of the answer sheet prior to taking the test. This process will take approximately 15 minutes.

STARTING AND COMPLETING THE EXAMINATION

You are not to open the examination booklet until instructed to do so by your proctor. If you complete the examination with more than 30 minutes remaining, you are free to leave after returning all examination materials to the proctor. Within 30 minutes of the end of the examination, you are required to remain until the end to avoid disruption to those still working and to permit orderly collection of all examination materials. Regardless of when you complete the examination, you are responsible for returning the numbered examination booklet assigned to you. Cooperate with the proctors collecting the examination materials. Nobody will be allowed to leave until the proctor has verified that all materials have been collected.

REFERENCES

The PE examination is open-book. Your board determines the reference materials and calculators that will be allowed. In general, you may use textbooks, handbooks, bound reference materials, and a non-communicating, battery-operated, silent, non-printing calculator. Calculating and computing devices having a QWERTY keypad arrangement similar to a typewriter or keyboard are not permitted, nor are communication devices such as pagers and cellular phones. States differ in their rules regarding calculators and references, and you should contact your board for specific advice.

SPECIAL ACCOMMODATIONS

If you require special accommodations in the test-taking procedure, you should communicate your need to your board office well in advance of the day of the examination so that appropriate arrangements may be determined.

UPDATES TO EXAMINATION INFORMATION

For updated exam specifications and design standards, errata for this book, an other information about exams, visit the NCEES Web site at www.ncees.org.

EXAMINATION SPECIFICATIONS

CIVIL BREADTH (AM) EXAMINATION
EFFECTIVE OCTOBER 2000

The civil engineering examination is a breadth and depth examination. This means that all examinees work the breadth (AM) exam and one of the five depth (PM) exams. The five areas covered in the civil engineering examination are environmental, geotechnical, structural, transportation, and water resources. The breadth exam contains questions from all five areas of civil engineering. The depth exams focus more closely on a single area of practice in civil engineering.

	Approximate Percentage of Examination
I. ENVIRONMENTAL	**20%**
A. Wastewater Treatment – wastewater flow rates, unit processes.	
B. Biology – toxicity, algae, stream degradation, temperature, disinfection, water taste & odor, BOD.	
C. Solid/Hazardous Waste – collection, storage/transfer, treatment, disposal, quantity estimates, site & haul economics.	
D. Ground Water and Well Fields – groundwater flow, aquifers (e.g., characterization).	
II. GEOTECHNICAL	**20%**
A. Subsurface Exploration & Sampling – drilling and sampling, soil classification, boring log interpretation, soil profile development.	
B. Engineering Properties of Soils – index properties, phase relationships, permeability.	
C. Soil Mechanics Analysis – pressure distribution, lateral earth pressure, consolidation, compaction.	
D. Shallow Foundations – bearing capacity, settlement, allowable bearing pressure.	
E. Earth Retaining Structures – gravity walls, cantilever walls, earth pressure diagrams, stability analysis.	
III. STRUCTURAL	**20%**
A. Loadings – dead & live loads, wind loads.	
B. Analysis – determinate analysis, shear diagrams, moment diagrams.	
C. Mechanics of Materials – flexure, shear, tension & compression, deflection.	
D. Materials – reinforced concrete, structural steel, timber, concrete mix design, masonry.	
E. Member Design – beams, slabs, columns, reinforced concrete footings, retaining walls, trusses.	
IV. TRANSPORTATION	**20%**
A. Traffic Analysis – capacity analysis.	
B. Construction – excavation/embankment, material handling, optimization, scheduling.	
C. Geometric Design – horizontal curves, vertical curves, sight distance.	
V. WATER RESOURCES	**20%**
A. Hydraulics – energy dissipation, energy/continuity equation, pressure conduit, open channel flow, flow rates, friction/minor losses, flow equations, hydraulic jump, culvert design, velocity control.	
B. Hydrology – storm characterization, storm frequency, hydrographs, rainfall intensity & duration, runoff analysis.	
C. Water Treatment – demands, hydraulic loading, storages (raw & treated water).	
TOTAL	**100%**

NOTES:

1. The knowledge areas specified as A, B, C, ... etc., are examples of kinds of knowledge, but they are not exclusive or exhaustive categories.
2. The breadth (AM) exam contains 40 multiple-choice questions. Examinee works all questions.

ENVIRONMENTAL DEPTH (PM) EXAMINATION
EFFECTIVE OCTOBER 2000

		Approximate Percentage of Examination
I.	**ENVIRONMENTAL**	**65%**
II.	**GEOTECHNICAL**	**10%**
III.	**WATER RESOURCES**	**25%**
	TOTAL	**100%**

I. ENVIRONMENTAL — **65%**

A. Wastewater Treatment
Wastewater flow rates, primary clarification, biological treatment, secondary clarification, chemical precipitation, sludge systems, digesters, disinfection, nitrification/denitrification, effluent limits, wetlands, unit processes, operations.

B. Biology (including micro & aquatic)
Toxicity, algae, food chain, stream degradation, organic load, oxygenation/ deoxygenation/ oxygen sag curve, eutrophication, temperature, indicator organisms, disinfection, water taste & odor, most probable number (MPN), BOD, quality control.

C. Solid/Hazardous Waste
Collection, storage/transfer, treatment, disposal, quantity estimates, site & haul economics, energy recovery, hazardous waste systems, applicable standards.

D. Ground Water and Well Fields
Dewatering, well analysis, water quality analysis, subdrain systems, groundwater flow, groundwater contamination, recharge, aquifers (e.g., characterization).

II. GEOTECHNICAL — **10%**

A. Subsurface Exploration and Sampling
Drilling and sampling procedures, soil classification, boring log interpretation, soil profile development.

B. Engineering Properties of Soils
Permeability.

C. Soil Mechanics Analysis
Compaction, seepage and erosion.

III. WATER RESOURCES — **25%**

A. Hydraulics
Energy/continuity equation, pressure conduit, open channel flow, detention/retention ponds, pump application and analysis, pipe network analysis, flow rates (domestic, irrigation, fire), surface water profile, cavitation, friction/minor losses, flow measurement devices, flow equations, culvert design, velocity control.

B. Hydrology
Storm characterization, storm frequency, hydrograph (unit & others), transpiration, evaporation, permeation, rainfall intensity & duration, runoff analysis, gauging stations, flood plain/floodway, sedimentation.

C. Water Treatment
Demands, hydraulic loading, storages (raw & treated water), rapid mixing, flocculation, sedimentation, filtration, disinfection, applicable standards.

TOTAL — **100%**

NOTES:

1. The knowledge areas specified as A, B, C, … etc., are examples of kinds of knowledge, but they are not exclusive or exhaustive categories.
2. Each depth (PM) exam contains 40 multiple-choice questions. Examinee chooses **one** depth exam and works all questions in the depth exam chosen.

GEOTECHNICAL DEPTH (PM) EXAMINATION
EFFECTIVE OCTOBER 2000

	Approximate Percentage of Examination
I. GEOTECHNICAL	**65%**
A. Subsurface Exploration and Sampling Drilling & sampling procedures, in-situ testing, soil classification, boring log interpretation, soil profile development.	
B. Engineering Properties of Soils Index properties, phase relationships, shear strength properties, permeability.	
C. Soil Mechanics Analysis Effective & total stresses, pore pressure, pressure distribution, lateral earth pressure, consolidation, compaction, slope stability, seepage and erosion.	
D. Shallow Foundations Bearing capacity, settlement, allowable bearing pressure, proportioning individual/combined footings, mat and raft foundations, pavement design.	
E. Deep Foundations Axial capacity (single pile/drilled shaft), lateral capacity (single pile/drilled shaft), settlement, lateral deflection, behavior of pile/drilled shaft groups, pile dynamics & pile load tests.	
F. Earth Retaining Structures Gravity walls, cantilever walls, mechanically stabilized earth wall, braced & anchored excavations, earth dams, earth pressure diagrams, stability analysis, serviceability requirements.	
G. Seismic Engineering Earthquake fundamentals, liquefaction potential evaluation.	
II. ENVIRONMENTAL	**10%**
A. Ground Water and Well Fields Dewatering, water quality analysis, groundwater contamination, aquifers (e.g., characterization).	
III. STRUCTURAL	**20%**
A. Loadings Dead & live loads, earthquake loads.	
B. Materials Concrete mix design.	
C. Member Design Reinforced concrete footings, pile foundations, retaining walls.	
IV. TRANSPORTATION	**5%**
A. Construction Excavation/embankment, pavement design.	
TOTAL	**100%**

NOTES:

1. The knowledge areas specified as A, B, C, ... etc., are examples of kinds of knowledge, but they are not exclusive or exhaustive categories.
2. Each depth (PM) exam contains 40 multiple-choice questions. Examinee chooses **one** depth exam and works all questions in the depth exam chosen.

STRUCTURAL DEPTH (PM) EXAMINATION
EFFECTIVE OCTOBER 2000

	Approximate Percentage of Examination
I. STRUCTURAL	**65%**

A. Loadings
Dead & live loads, moving loads, wind loads, earthquake loads, repeated loads.

B. Analysis
Determinate, indeterminate, shear diagrams, moment diagrams.

C. Mechanics of Materials
Flexure, shear, torsion, tension & compression, combined stresses, deflection.

D. Materials
Reinforced concrete, pre-stressed concrete, structural steel, timber, concrete mix design, masonry, composite construction.

E. Member Design
Beams, slabs, columns, reinforced concrete footings, pile foundations, retaining walls, trusses, braces & connections, shear and bearing walls.

F. Failure Analysis
Buckling, fatigue, failure modes.

G. Design Criteria
UBC, BOCA, SBC, ACI, PCI, AISC, NDS, AASHTO, ASCE-7

II. GEOTECHNICAL — **25%**

A. Subsurface Exploration and Sampling
Boring log interpretation.

B. Soil Mechanics Analysis
Pressure distribution, lateral earth pressure.

C. Shallow Foundations
Bearing capacity, settlement, proportioning individual/combined footings, mat & raft foundations

D. Deep Foundations
Axial capacity - Single pile/drilled shaft.
Lateral capacity - Single pile/drilled shaft.
Behavior of pile/drilled shaft groups.

E. Earth Retaining Structures
Gravity walls, cantilever walls, braced & anchored excavations, earth pressure diagrams, stability analysis.

III. TRANSPORTATION — **10%**

A. Construction
Excavation/embankment, material handling, optimization, scheduling.

TOTAL — **100%**

NOTES:

1. The knowledge areas specified as A, B, C, ... etc., are examples of kinds of knowledge, but they are not exclusive or exhaustive categories.
2. Each depth (PM) exam contains 40 multiple-choice questions. Examinee chooses **one** depth exam and works all questions in the depth exam chosen.

TRANSPORTATION DEPTH (PM) EXAMINATION
EFFECTIVE OCTOBER 2000

Approximate Percentage of Examination

I. TRANSPORTATION **65%**

A. Traffic Analysis
Traffic signal, speed studies, capacity analysis, intersection analysis, parking operations, traffic volume studies, mass transit studies, sight distance, traffic control devices, pedestrian facilities, bicycle facilities, driver behavior/performance.

B. Transportation Planning
Origin-destination studies, site impact analysis, capacity analysis, optimization/cost analysis, trip generation/distribution/assignment.

C. Construction
Excavation/embankment, material handling, optimization, scheduling, mass diagrams, pavement design.

D. Geometric Design
Horizontal curves, vertical curves, sight distance, superelevation, vertical/horizontal clearances, acceleration & deceleration, intersections/interchanges.

E. Traffic Safety
Accident analysis, roadside clearance analysis, counter-measurement development, economic analysis, conflict analysis.

II. GEOTECHNICAL **15%**

A. Subsurface Exploration and Sampling
Soil classification, boring log interpretation, soil profile development.

B. Engineering Properties of Soils
Index properties, phase relationships.

C. Soil Mechanics Analysis
Compaction, seepage & erosion.

D. Shallow Foundations
Pavement design.

III. WATER RESOURCES **20%**

A. Hydraulics
Open channel flow, flow rates (domestic, irrigation, fire), flow equations, culvert design, velocity control.

B. Hydrology
Rainfall intensity & duration, runoff analysis, flood plain/floodway.

TOTAL **100%**

NOTES:

1. The knowledge areas specified as A, B, C, ... etc., are examples of kinds of knowledge, but they are not exclusive or exhaustive categories.
2. Each depth (PM) exam contains 40 multiple-choice questions. Examinee chooses **one** depth exam and works all questions in the depth exam chosen.

WATER RESOURCES DEPTH (PM) EXAMINATION
EFFECTIVE OCTOBER 2000

	Approximate Percentage of Examination
I. WATER RESOURCES	**65%**
II. ENVIRONMENTAL	**25%**
III. GEOTECHNICAL	**10%**
TOTAL	**100%**

I. WATER RESOURCES — 65%

A. Hydraulics
Spillway capacity, energy dissipation, energy/continuity equation, pressure conduit, open channel flow, detention/retention ponds, pump application and analysis, pipe network analysis, stormwater collection, flow rates (domestic, irrigation, fire), surface water profile, cavitation, friction/minor losses, sub- & supercritical flow, hydraulic jump, flow measurement devices, flow equations, culvert design, velocity control.

B. Hydrology
Storm characterization, storm frequency, hydrographs (unit & others), transpiration, evaporation, permeation, rainfall intensity & duration, runoff analysis, gauging stations, flood plain/floodway, sedimentation.

C. Water Treatment
Demands, hydraulic loading, storages (raw & treated water), rapid mixing, flocculation, sedimentation, filtration, disinfection, applicable standards.

II. ENVIRONMENTAL — 25%

A. Wastewater Treatment
Unit processes

B. Biology (including micro & aquatic)
Toxicity, algae, food chain, stream degradation, organic load, eutrophication, temperature, indicator organisms, disinfection, water taste & odor, most probable number (MPN), BOD, quality control.

C. Ground Water and Well Fields
Well analysis, water quality analysis, groundwater flow, groundwater contamination, recharge, aquifers (e.g., characterization).

III. GEOTECHNICAL — 10%

A. Subsurface Exploration and Sampling
Soil classification, boring log interpretation, soil profile development.

B. Engineering Properties of Soils
Permeability.

C. Soil Mechanics Analysis
Seepage and erosion.

TOTAL — 100%

NOTES:

1. The knowledge areas specified as A, B, C, ... etc., are examples of kinds of knowledge, but they are not exclusive or exhaustive categories.
2. Each depth (PM) exam contains 40 multiple-choice questions. Examinee chooses **one** depth exam and works all questions in the depth exam chosen.

CIVIL BREADTH
MORNING SAMPLE QUESTIONS

This book contains 20 civil breadth questions, half the number on the actual exam

CIVIL MORNING SAMPLE QUESTIONS

101. A stabilization pond treats a flow of 1.0 MGD.

Existing conditions are:

Winter liquid temperature = 15°C
Overall first-order removal rate BOD
removal rate constant (k) = 0.25 d^{-1} at 20°C
θ = 1.06 (temperature coefficient)

Assuming a dispersal factor d of 0.5, the minimum pond volume (MGAL) under winter conditions (15°C) to achieve 80% removal of BOD is most nearly:

(A) 10
(B) 14
(C) 18
(D) 22

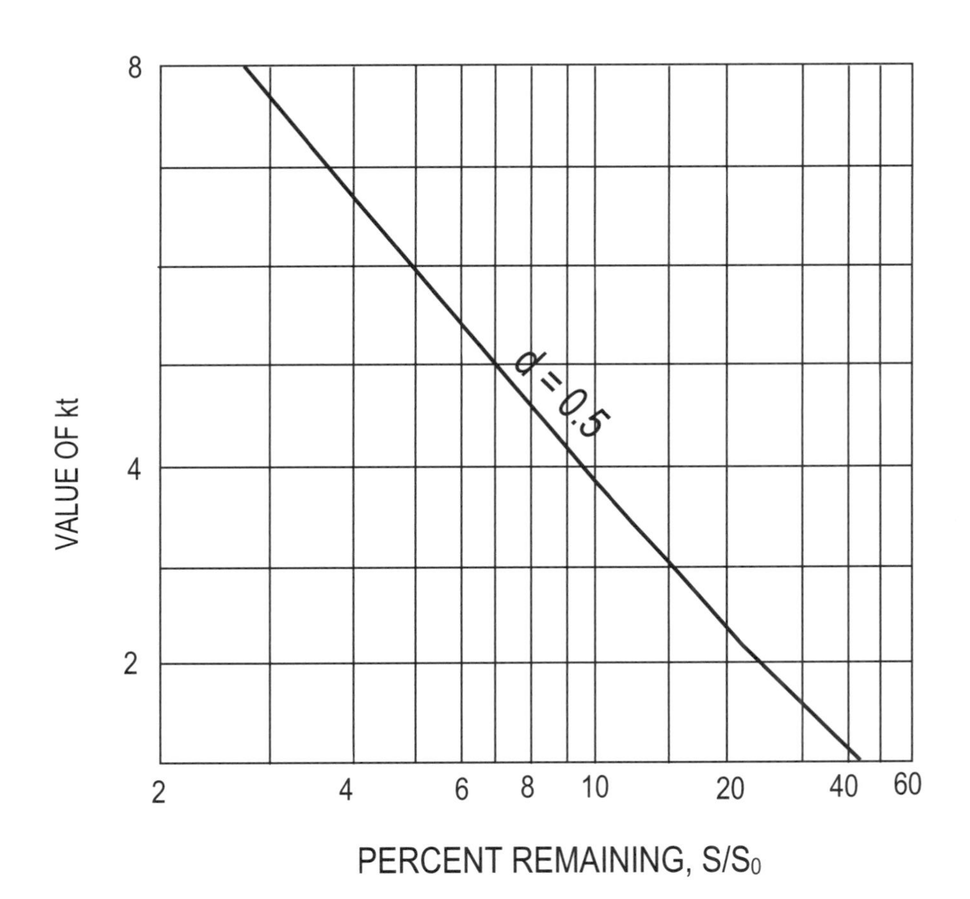

GO ON TO THE NEXT PAGE

102. A water treatment plant designed to treat surface water consists of coagulation/flocculation, sedimentation, filtration, activated carbon adsorption, and chlorine disinfection treatment processes. The most effective treatment process for removing tastes and odors is:

(A) coagulation/flocculation
(B) filtration
(C) activated carbon adsorption
(D) chlorine disinfection

103. All the municipal solid waste (MSW) generated by a community is treated at a waste-to-energy (WTE) mass burner facility. In response to a new state law, the local solid-waste management authority is instituting a curbside recycling program for residents. Given the recycling goals, the impacts on the amount of electricity generated at the WTE facility are to be calculated. **Table 1** shows the composition of the MSW feed to the facility and its design energy content.

The proposed curbside recycling program will divert the components shown in **Table 2** from the WTE facility feed stream. The table shows the percentage of the component that is recycled.

The reduction in energy content of the waste after the recycling program is implemented is most nearly:

(A) 30%
(B) 40%
(C) 50%
(D) 60%

GO ON TO THE NEXT PAGE

TABLE 1

MSW Composition (As Collected)		
Component	Percent by Weight (As Collected)	Energy Content (Design Basis) Btu/lb
Organic		
Food Wastes	8.0	2,500
Paper	39.0	6,700
Cardboard	6.0	7,000
Plastics	5.0	15,000
Textiles	2.5	7,200
Rubber	1.0	10,500
Leather	0.3	7,500
Yard Wastes	20.5	3,000
Wood	3.0	8,000
Inorganic		
Glass	6.0	75
Metals	5.7	200
Dirt, ash, etc.	3.0	3,000
Total	100.0	

TABLE 2

MSW Component Recycling Percentage	
Component	% Recycling
Paper	50.0
Cardboard	75.0
Plastics	25.0
Yard Wastes	25.0
Wood	25.0

GO ON TO THE NEXT PAGE

104. Groundwater monitoring wells have been installed at the site of a proposed sanitary landfill such that Well B is located 1,500 feet north and 300 feet west of Well A.

A proposed containment cell, having bottom dimensions of 500 feet (north to south) and 100 feet (east to west), is to be located such that the southeast corner is 100 feet west and 500 feet north of Well A.

The bottom of the landfill cell must be a minimum of 5 feet above the groundwater elevation. The elevations that were determined at each well location are shown in the table below.

The minimum bottom elevation (ft) for the proposed landfill cell if the bottom is to be level is most nearly:

(A) 227
(B) 232
(C) 237
(D) 242

Well	Groundwater Elevation (ft)	Ground Surface Elevation (ft)
A	229.75	248.75
B	222.25	243.75

GO ON TO THE NEXT PAGE

105. The effective overburden pressure (psf) at the middle of the clay layer shown in the subsurface section below is most nearly:

(A) 1,001
(B) 1,076
(C) 1,388
(D) 1,700

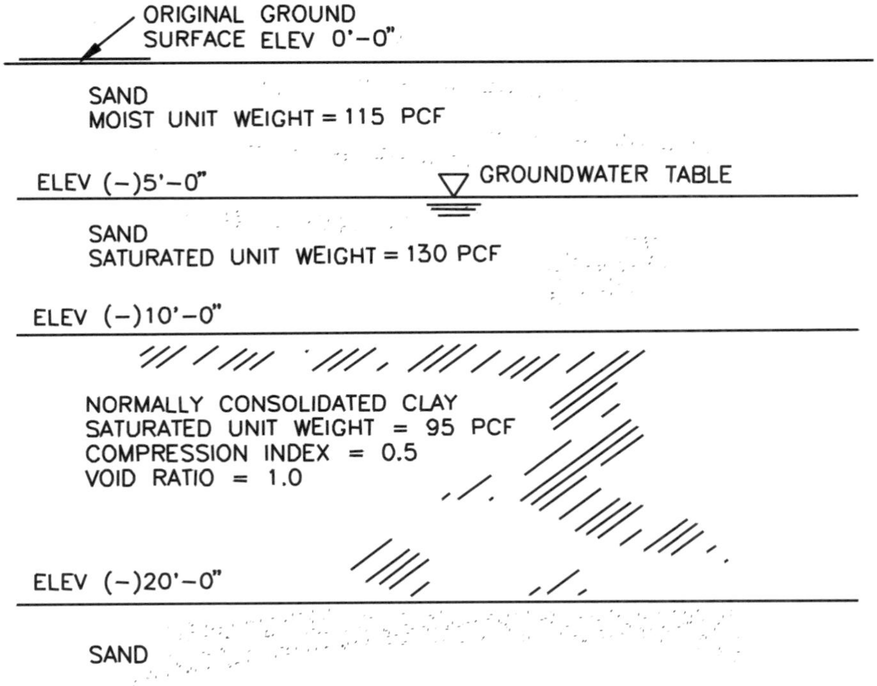

ORIGINAL GROUND
SURFACE ELEV 0'-0"
SAND
MOIST UNIT WEIGHT = 115 PCF
ELEV (-)5'-0"
GROUNDWATER TABLE
SAND
SATURATED UNIT WEIGHT = 130 PCF
ELEV (-)10'-0"
NORMALLY CONSOLIDATED CLAY
SATURATED UNIT WEIGHT = 95 PCF
COMPRESSION INDEX = 0.5
VOID RATIO = 1.0
ELEV (-)20'-0"
SAND

SUBSURFACE SECTION
NOT TO SCALE

106. Given a sand sample with a dry density equal to 107 pcf and specific gravity of soil solids equal to 2.65, the void ratio of the sand is most nearly:

(A) 0.3
(B) 0.5
(C) 0.8
(D) 1.0

107. Refer to the following data:

Dry unit weight of soil in borrow pit	87.0 pcf
Moisture content in borrow pit	13.0%
Specific gravity of the soil particles	2.70
Modified Proctor optimum moisture content	14.0%
Modified Proctor maximum dry density	116.0 pcf

Assume that soil from a borrow pit is transported to a construction site to construct 500,000 yd^3 of compacted roadway embankment. Due to handling and evaporation, the soil arrives at the construction site with the moisture content equal to 9%. The soil is placed and compacted to 90% of the Modified Proctor maximum dry density. The total volume of water (gallons) that must be added to the soil to increase the moisture content to the optimum level is most nearly:

(A) 313,000
(B) 4,500,000
(C) 8,500,000
(D) 9,400,000

108. A retaining wall will be constructed at a site. Assume the sand backfill has a friction angle of 28°. According to Rankine's theory, the active lateral earth pressure coefficient is most nearly:

(A) 0.30
(B) 0.35
(C) 0.40
(D) 0.50

GO ON TO THE NEXT PAGE

109. Codes:
ACI 530-95, *Building Code Requirements for Masonry Structures*

Materials:
Hollow concrete masonry units f'_m = 1,500 psi with Type S mortar. Cells with reinf. grouted
Steel reinforcement ASTM A615 Grade 60

Loads:
Roof dead load = 15 psf
Non-reducible roof snow load = 40 psf
Average wall dead load = 54 psf
Design wind (pressure or suction) = 20 psf
Seismic forces do not govern.

The total axial load, P, in pounds per linear foot at the midheight of the wall is most nearly:

(A) 60
(B) 980
(C) 1,310
(D) 1,970

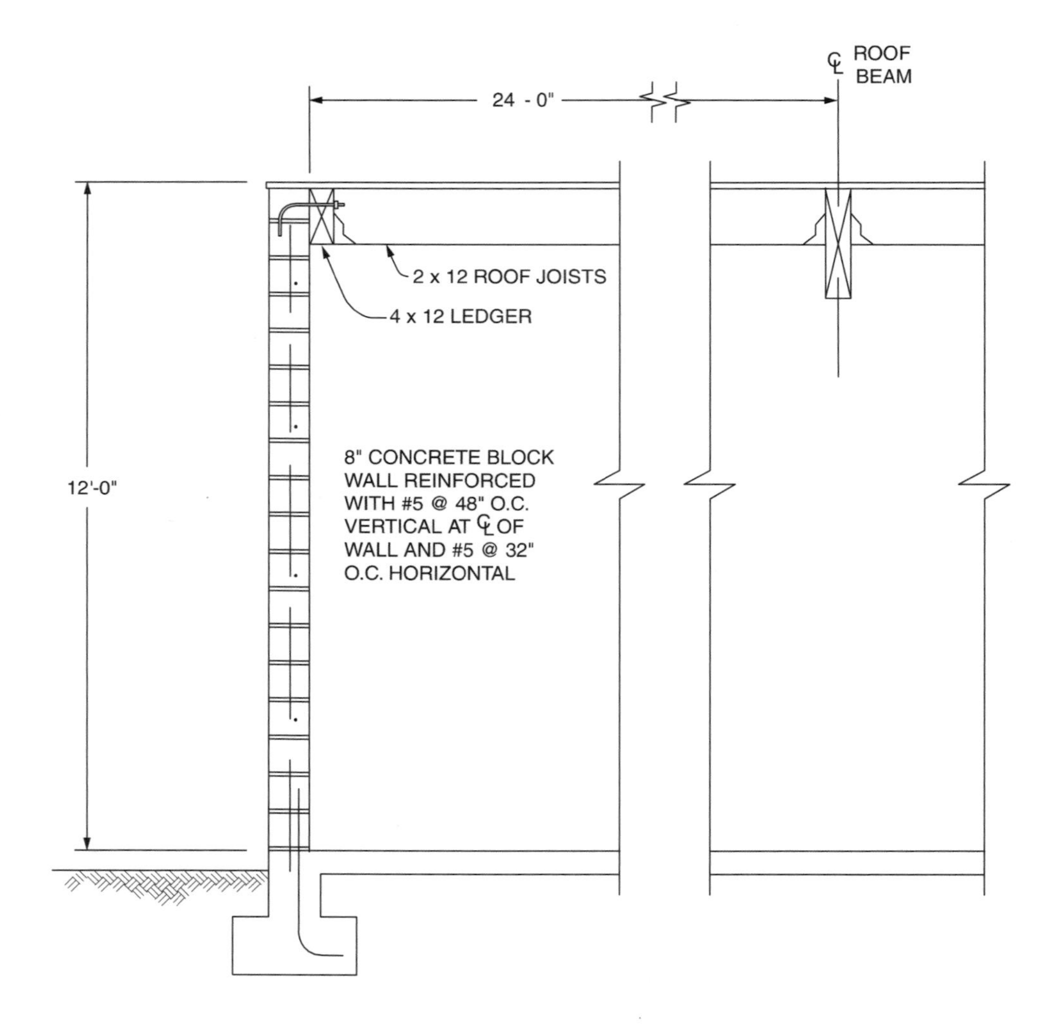

110. Assume that the basement wall shown below is supported by the first floor and basement footing. For an equivalent fluid pressure of 40 pcf, the maximum wall bending moment (ft-lb/ft) is most nearly:

(A) 200
(B) 340
(C) 500
(D) 1,600

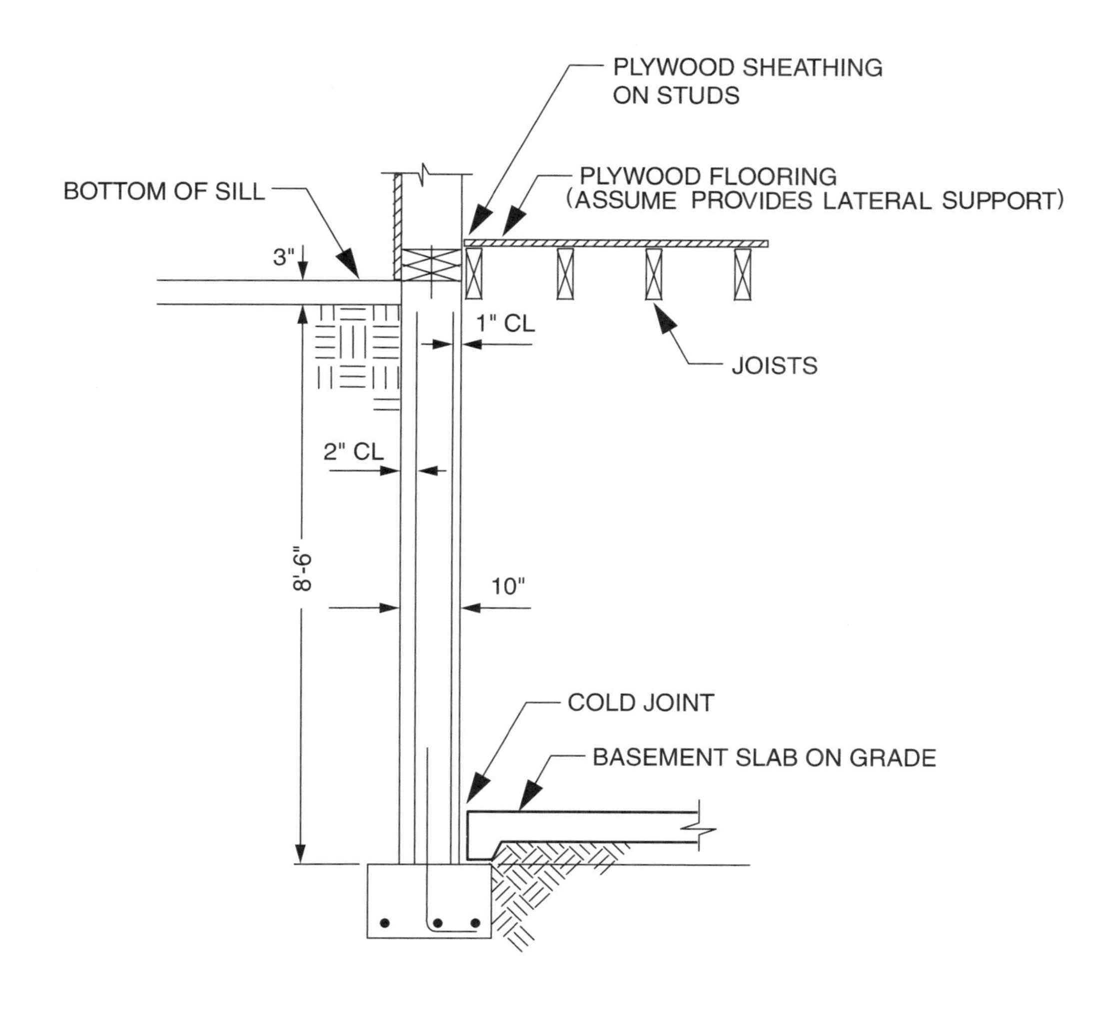

GO ON TO THE NEXT PAGE

For **Questions 111 and 112**, assume the depth to reinforcement d = 42 inches (from extreme compression fiber to centroid of tension reinforcement) and that the girder span is 40'-0".

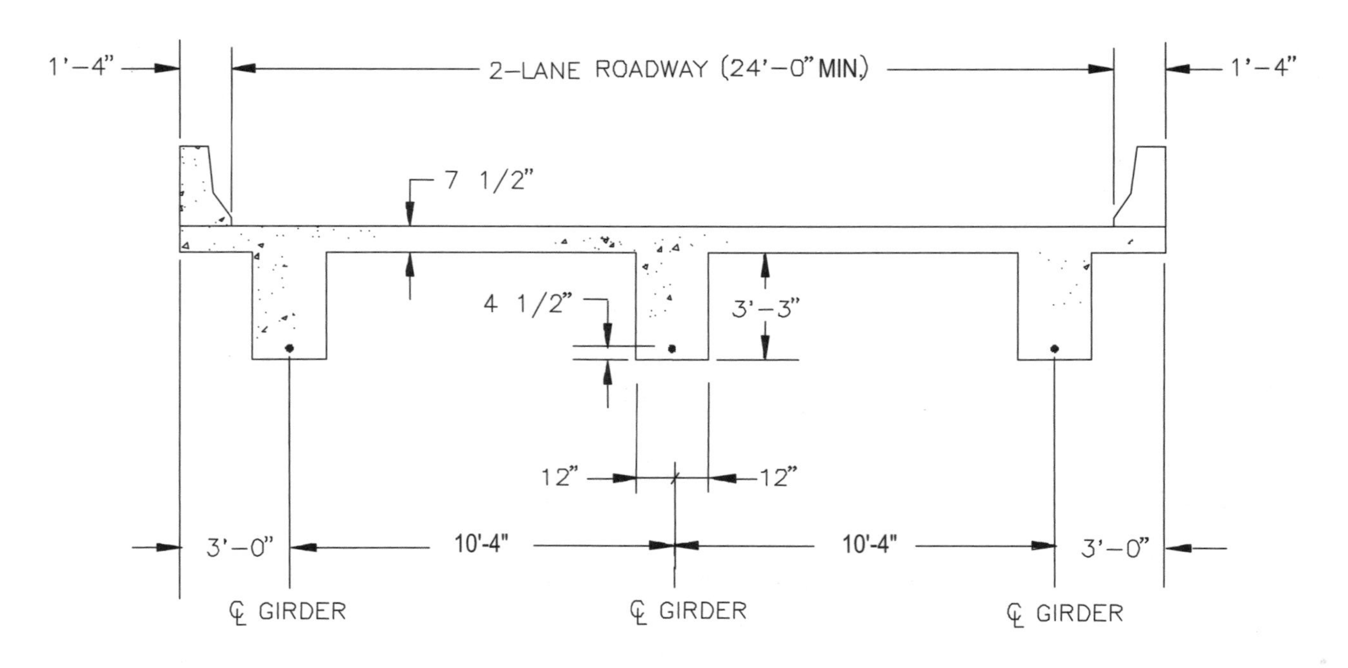

BRIDGE CROSS–SECTION
NOT TO SCALE

The bridge cross-section is for a simple-span bridge having three reinforced-concrete T-girders.

Design Data:

Normal weight concrete, f'_c	=	5,000 psi
Reinforcement bars, F_y	=	60,000 psi
Live Load	=	H15-44

Use AASHTO *Standard Specifications for Highway Bridges*, 16th edition (1996).

111. For the center girder, assuming a factored moment M_u = 2,000 ft-kips, the minimum area (in^2) of reinforcement needed to resist the moment is most nearly:

(A) 7.5
(B) 11
(C) 15
(D) 19

112. The nominal shear strength V_c (kips) provided by the concrete is most nearly:

(A) 72.5
(B) 104
(C) 125
(D) 145

GO ON TO THE NEXT PAGE

113. Given an eight-lane freeway with free-flow speed = 70 mph, the maximum service flow rate per lane at level of service D (pcphpl) is most nearly:

(A) 1,855
(B) 1,980
(C) 2,045
(D) 2,400

114. A highway contractor has located a stone deposit that can conveniently be quarried. It is estimated that 50,000 yd^3 per year can be produced for 10 years. This would exhaust the supply of stone. Cleanup costs equal the value of the land at that time. If purchased from others, crushed rock would cost $7.00/$yd^3$ delivered to the job site. If quarried and crushed by the contractor, crushed rock would cost $5.50/$yd^3$ (including all equipment, labor, and supplies) delivered to the job site. The minimum attractive rate of return (MARR) is 10%.

The maximum amount the contractor can afford to pay for this deposit without a financial loss is most nearly:

(A) $284,300
(B) $424,800
(C) $460,900
(D) $481,400

115. The station of the P.I. is most nearly:

(A) 29+94.90
(B) 33+42.84
(C) 34+77.82
(D) 35+10.47

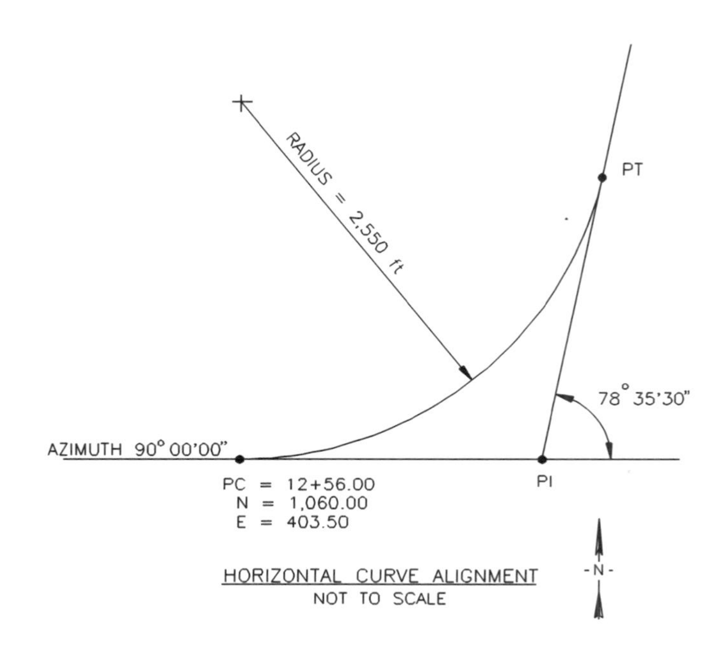

116. The tangent vertical alignment of a section of proposed highway is shown in the figure below. The horizontal distance (feet) between P.V.I.$_1$ and P.V.I.$_2$ is most nearly:

(A) 3,395
(B) 3,400
(C) 3,410
(D) 3,425

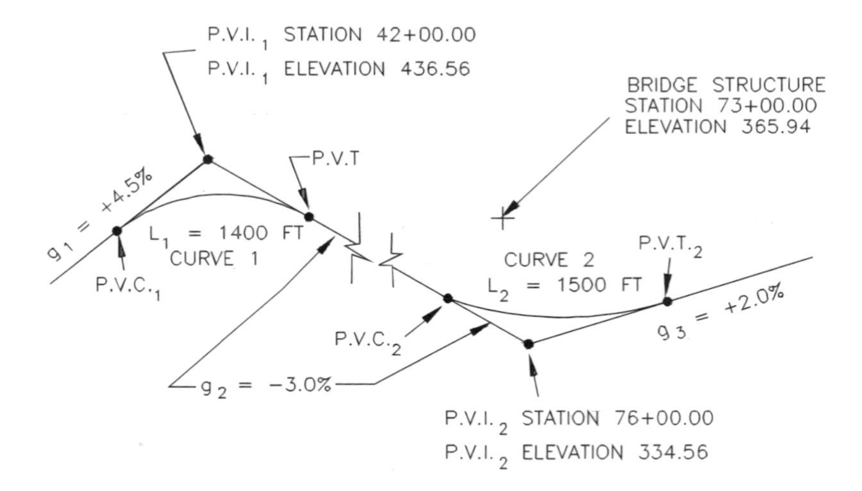

PROPOSED HIGHWAY VERTICAL ALIGNMENT
NOT TO SCALE

117. The best estimate for the Hazen-Williams resistance coefficient for a new 10-inch-diameter ductile iron pipe is most nearly:

(A) 150
(B) 130
(C) 100
(D) 80

118. A pump in the figure below is to deliver water through the existing system indicated in the figure.

Which of the following will reduce the tendency for pump cavitation?

I. Increasing the discharge pipe diameter
II. Lowering the pump elevation
III. Increasing the suction diameter

(A) II and III only
(B) II only
(C) I and II only
(D) I, II , and III

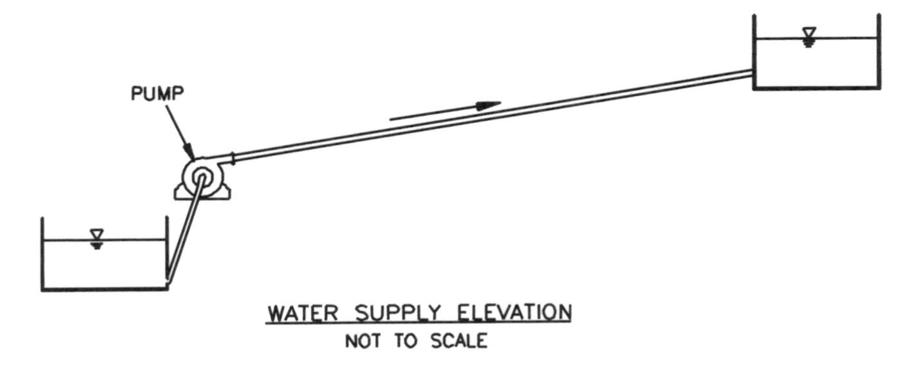

119. In general, converting an area from natural grassland to 1/4-acre-lot single family housing will have what effect on time of concentration and amount of runoff?

(A) Decrease time of concentration and increase runoff.
(B) Decrease time of concentration and runoff.
(C) Increase time of concentration and decrease runoff.
(D) Increase time of concentration and runoff.

GO ON TO THE NEXT PAGE

120. The results of a well water analysis are given below:

Ca^{++}	51 mg/L
Mg^{++}	12 mg/L
Na^{+}	25 mg/L
$SO_4^{=}$	65 mg/L
Cl^{-}	25 mg/L
F^{-}	0.4 mg/L
NO_3^{-}	14 mg/L as NO_3^{-}–N
pH	7.8
H_2S	3.4 mg/L as S
Alkalinity	84 mg/L as $CaCO_3$
Total coliforms	MPN 2.2 org/100 ml
Turbidity	6.2 NTU
Chlorine demand	9.1 mg/L
TDS	332 mg/L
Temperature	25°C

The total hardness (mg/L as $CaCO_3$) is most nearly:

(A) 63
(B) 88
(C) 156
(D) 177

CIVIL DEPTH
AFTERNOON SAMPLE QUESTIONS

ENVIRONMENTAL
AFTERNOON SAMPLE QUESTIONS

This book contains 20 environmental depth questions, half the number on the actual exam.

In some cases the same question appears in more than one of the depth modules because there is crossover of knowledge between the depth areas of civil engineering. This use of the same question on different depth modules could occur on the examination.

501. A city wants to design a sludge dewatering system for their 20-MGD secondary, activated sludge treatment plant shown below.

The sludge volume reduction (%) achieved by the thickener is most nearly:

(A) 50
(B) 58
(C) 65
(D) 70

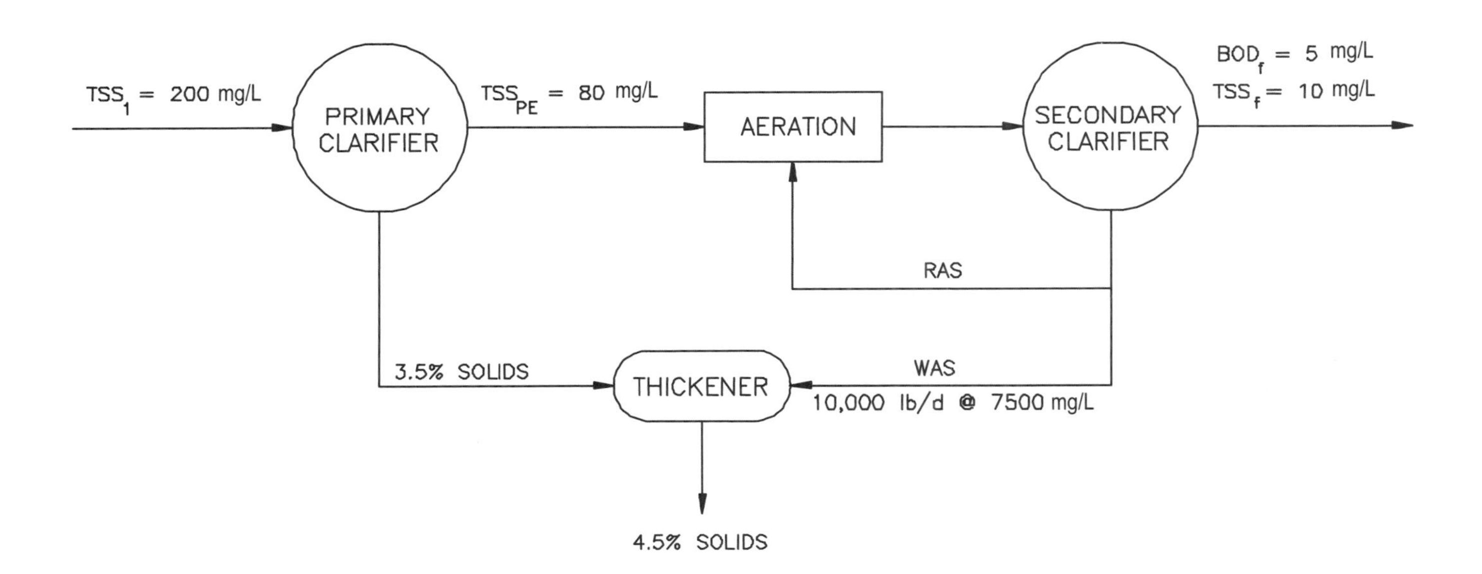

502. A city produces 15,000 gpd of thickened municipal primary waste sludge containing 10,000 lb of solids per day. The solids are 70% volatile. This sludge is digested in a high-rate anaerobic digester for 15 days (hydraulic detention time). The maximum loading of the digester is 100 lb volatile solids per 1,000 ft^3. Volatile solids reduction is 60%.

The total solids leaving the digester per day per unit volume of the digester [lb/(ft^3-day)] is most nearly:

(A) 0.06
(B) 0.08
(C) 0.14
(D) 0.20

503. A completely mixed aeration tank treats effluent from a primary clarifier in an activated sludge plant. The raw wastewater to the primary clarifier contains 275 mg/L soluble BOD_5, and the plant is designed to achieve 95% removal of BOD_5. The design sludge age is selected to be 10 days, K_d is 0.1 day^{-1}, and maximum cell yield (Y) is 0.5 mg VSS/mg BOD_5, and the flow is 2.0 MGD. The BOD_5 removal efficiency of the primary clarifier is 30%. Mixed liquor volatile suspended solids in the aerator tank is to be 3,000 mg/L.

The volume of the aeration tank (gal) necessary to meet all of the above conditions is most nearly:

(A) 115,000
(B) 298,000
(C) 435,000
(D) 596,000

GO ON TO THE NEXT PAGE

504. A conventional activated sludge plant treats 2.0 MGD as shown in the figure below. The raw wastewater contains 250 mg/L suspended solids. The mixed liquor suspended solids (MLSS) is 2,500 mg/L. The recycled sludge flow is 800,000 gpd containing 8,000 mg/L suspended solids. Suspended solids removal efficiency of the primary clarifier is 50%, and the primary sludge contains 4% suspended solids. Effluent solids content = 25 mg/L.

The primary sludge and waste sludge flow rates (gpd) are, respectively, most nearly:

(A) 6,200 and 6,900
(B) 8,900 and 69,000
(C) 6,200 and 69,000
(D) 6,200 and 620,000

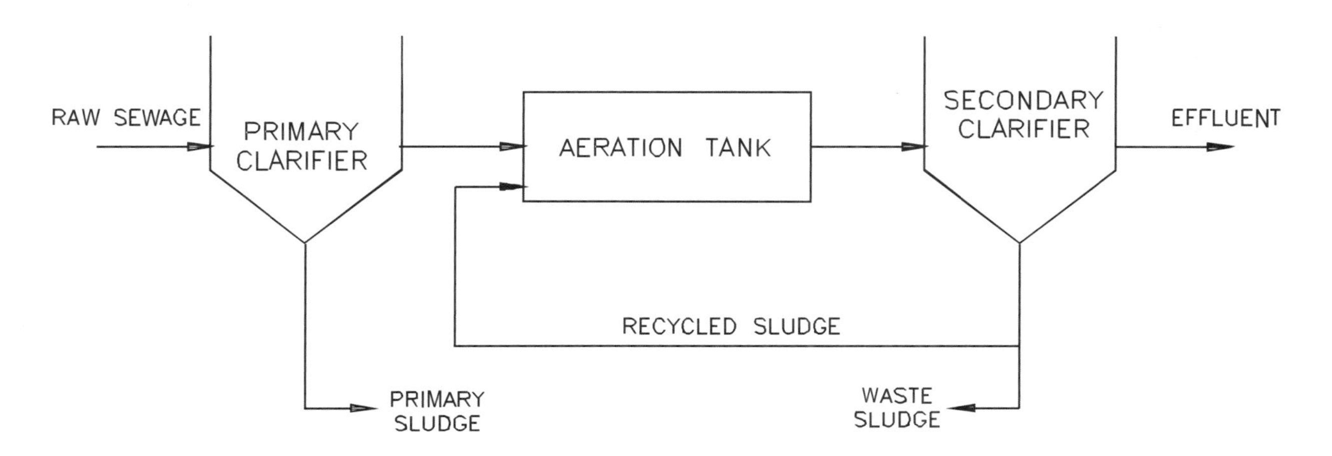

ACTIVATED SLUDGE PLANT – SCHEMATIC

505. You are an engineer working on sewer designs for a new residential subdivision being added to an existing city. The sanitary sewer servicing the subdivision will pick up the wastewater flow from a population of 5,775 new people. Annual average wastewater flow is estimated at 110 gpd, per capita.

It is proposed by the developer that this subdivision wastewater flow be put into an existing gravity sewer main nearby rather than paying for a new force main to the treatment plant. The existing sewer main is a 36-inch reinforced concrete pipe with a Manning's "n" = 0.017. Its slope is 0.002 ft/ft. The Public Works Department has determined the existing peak flow in the line is 7.0 MGD.

For all calculations, assume n is constant for all depths of flow. A peaking factor of 3.8 should be used.

The depth of flow (inches) in the existing sewer main under peak flow conditions after connecting the subdivision will be most nearly:

(A) 15
(B) 21
(C) 23
(D) 28

506. A secondary treated effluent from a 4.0-MGD wastewater treatment plant is discharged into a receiving stream.

The wastewater has a BOD_5 of 20 mg/L. The receiving stream upstream from the point of wastewater discharge has a flow rate of 18 cfs and a BOD_5 of 4.0 mg/L.

The BOD reaction rate constant is estimated at 0.23 day^{-1} (base e at 20°C). Reaeration and deoxygenation are the only major factors affecting the dissolved oxygen concentration in the stream after mixing with wastewater effluent.

The ultimate BOD (mg/L) just downstream of the effluent discharge into the receiving stream is most nearly:

(A) 4
(B) 8
(C) 12
(D) 16

GO ON TO THE NEXT PAGE

507. Surface waters having a significant concentration of algae tend to exhibit the following characteristics during daylight:

(A) increased DO concentration
(B) decreased pH
(C) increased CO_2 concentration
(D) decreased DO concentration

508. Assume the following characteristics of the wastewater/river mixture at the point of discharge:

Dissolved oxygen concentration	6 mg/L
Ultimate BOD	10 mg/L
Temperature	20°C
River velocity	1 fps
Reaeration rate constant (base e)	0.40 day^{-1}
Deoxygenation rate constant (base e)	0.23 day^{-1}

Assume that no other wastewater sources are discharged into the river.

The location (miles) of the critical dissolved oxygen concentration from the point of wastewater discharge is most nearly:

(A) 3
(B) 11
(C) 28
(D) 76

GO ON TO THE NEXT PAGE

509. The toxicity of ammonia nitrogen to fish is increased by:

(A) increasing pH
(B) decreasing temperature
(C) increasing ionic strength
(D) decreasing pH

510. A BOD sample was incubated at 27°C for 7 days and was found to measure 100 mg/L. Given $k_{20} = 0.23\ d^{-1}$, base e. The concentration of BOD (mg/L) for this sample that would be measured at an incubation time of 5 days and a temperature of 20°C would be most nearly:

(A) 125
(B) 76
(C) 35
(D) 20

511. Assume a waste-to-soil cover volume ratio of 10:1 for a municipal solid-waste landfill (MSWLF). If the average unit density of the waste as delivered to the landfill is 300 lbm/yd^3 and the compaction ratio at the landfill is 0.30, the volume of soil (yd^3 in place) required to cover 6,000 tons of the incoming raw waste is most nearly:

(A) 13,333
(B) 3,600
(C) 1,320
(D) 1,200

GO ON TO THE NEXT PAGE

512. You are to evaluate the impact of moisture in a municipality's solid waste stream on their processing and disposal systems. The table below is an analysis of the municipality's waste stream based on records for the past 3 years. All values are expressed as percentages by weight. The moisture content of each component is relatively constant.

The moisture content in the average waste stream is most nearly:

(A) 36%
(B) 30%
(C) 25%
(D) 21%

Solid Waste Characteristics		
Component	**Stream Composition, Percent by Weight Average**	**Average Moisture, Percent by Weight**
Organic		
Food	8.0	65.0
Paper	39.0	7.0
Cardboard	6.0	6.0
Plastic	5.0	2.5
Cloth	2.5	10.5
Rubber	1.0	2.5
Leather	0.3	10.0
Grass, Plants, etc.	20.5	55.0
Wood	3.0	27.5
Inorganic		
Glass	6.0	2.5
Metals	5.7	3.0
Dirt, Ash, etc.	3.0	9.0
	100%	

GO ON TO THE NEXT PAGE

513. A manufacturer has been ordered by the Environmental Protection Agency to clean up the contaminated groundwater underneath its plant. The concentrations of the various contaminants in the groundwater are shown in the table below.

The manufacturer has decided to pump the groundwater to the surface at the rate of 100 gpm and remove the contaminants by various treatment processes before discharging the water to Little Creek. The flow in Little Creek is fairly constant at 1 cfs. The upstream concentration of the contaminants of interest and the allowable downstream limits are shown in the figure below. The downstream limits were set by the Environmental Protection Agency.

The removal efficiencies (%) that should be achieved for tetrachloroethylene, toluene, chromium, and lead are most nearly:

(A) 70, 85, 70 and 40
(B) 85, 70, 85 and 40
(C) 97, 35, 97 and 40
(D) 97, 35, 97 and 0

Concentration of Groundwater Pollutants	
Contaminant	**Concentration (μg/L)**
Tetrachloroethylene	165
Toluene	7,300
Chromium (Hexavalent)	3,800
Lead	200

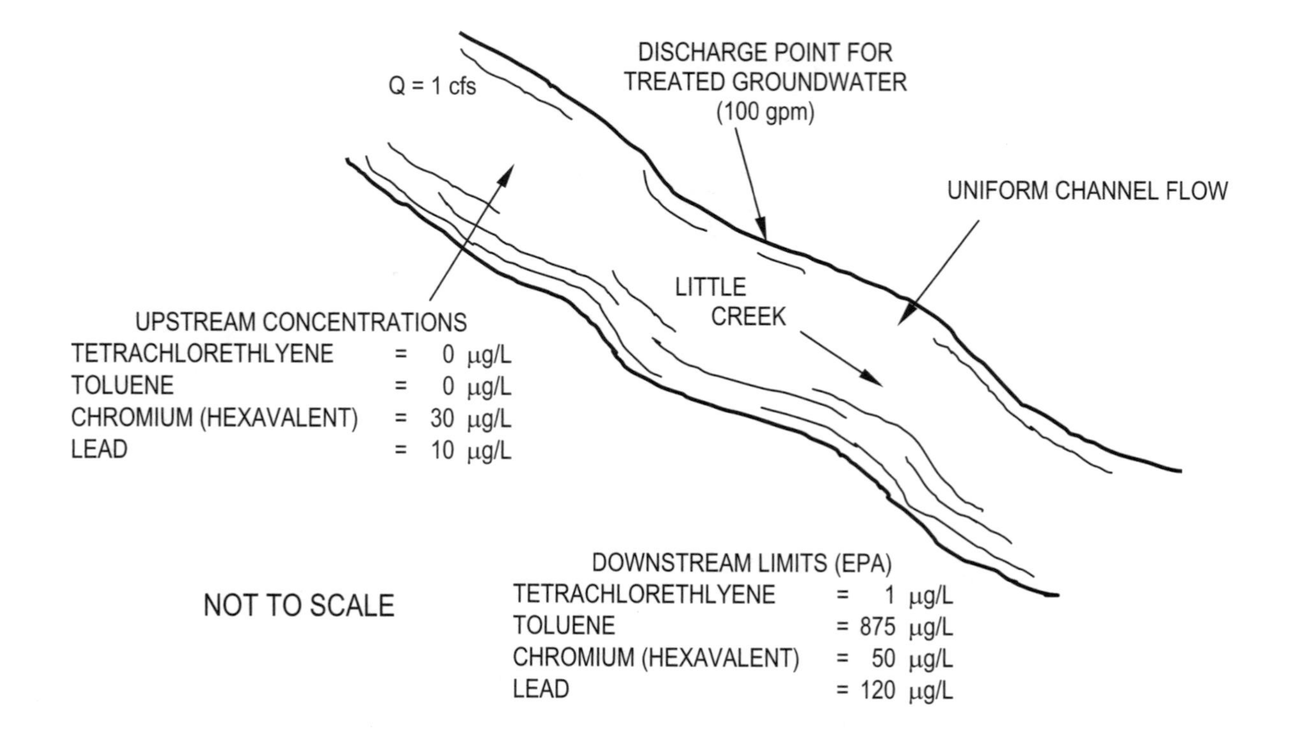

Environmental Question 514 also appears as Water Resources Question 519.

514. A soil sample was tested in the apparatus as shown in the following figure.

The coefficient of permeability (inches per minute) of the soil sample is most nearly:

(A) 0.02
(B) 0.14
(C) 0.21
(D) 0.44

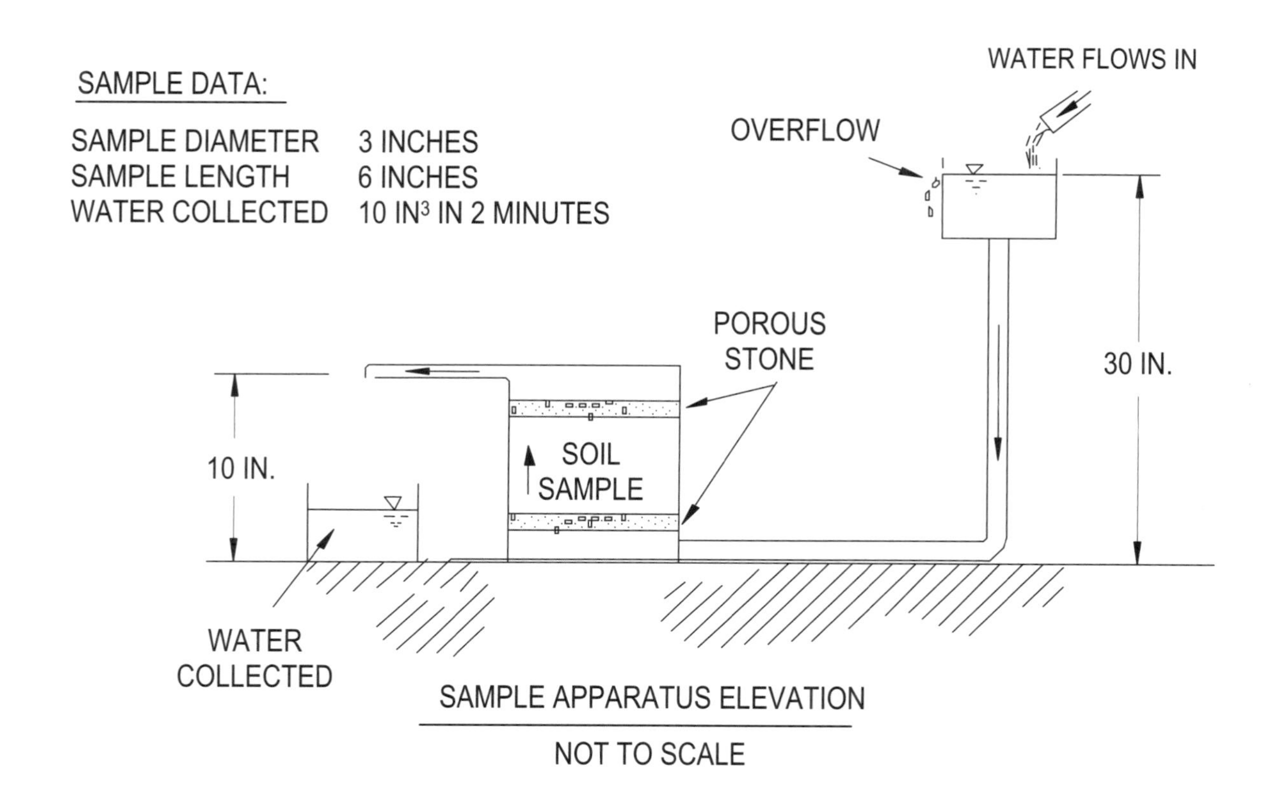

GO ON TO THE NEXT PAGE

Environmental Question 515 also appears as Water Resources Question 520.

515. A soil sample was tested in the apparatus as shown in the following figure.

The coefficient of permeability (inches per minute) of the soil sample is most nearly:

(A) 0.010
(B) 0.013
(C) 0.018
(D) 0.023

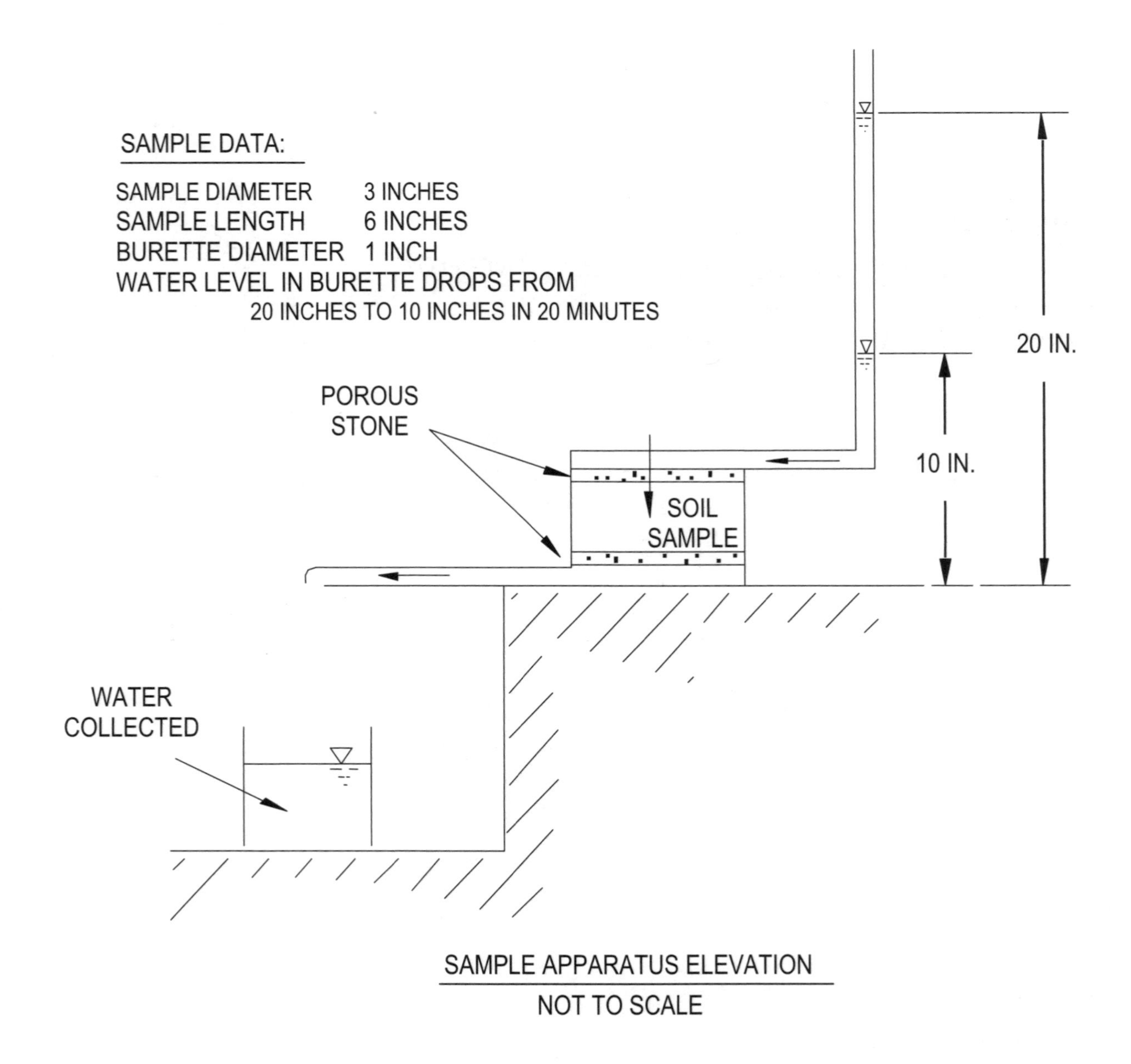

SAMPLE APPARATUS ELEVATION
NOT TO SCALE

516. The pump shown in **Figure 1** with the characteristics given in **Figure 2** is to deliver water through the existing system indicated in the figure. The Hazen-Williams equation is to be used to estimate friction losses. The total length of pipe is 3,000 feet, and the pipe is made of cast iron with a diameter of 8 inches and a Hazen-Williams coefficient of 100. Elevations are indicated on **Figure 1**.

The efficiency of the pump, at the point of operation, is most nearly:

(A) 55%
(B) 66%
(C) 71%
(D) 79%

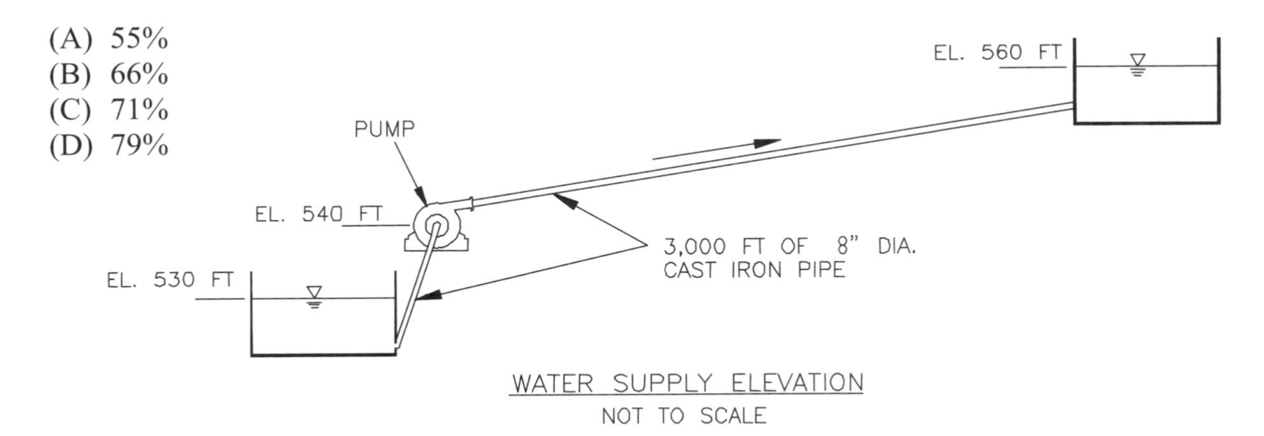

FIGURE 1

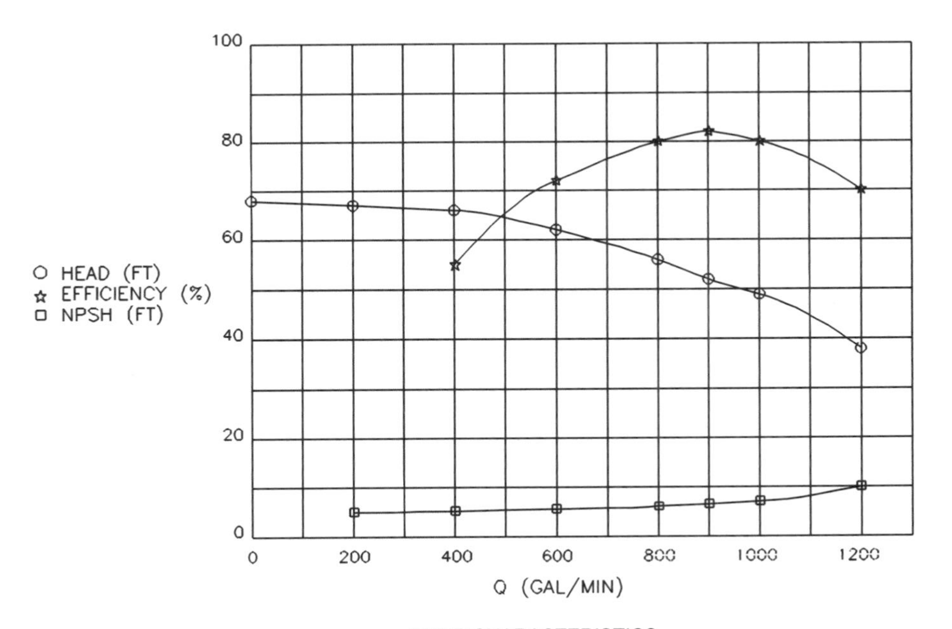

PUMP CHARACTERISTICS

FIGURE 2

GO ON TO THE NEXT PAGE

517. If the depth of flowing water in a ditch with Manning's roughness of 0.02 and an average bed slope of 0.5% is 2 feet, the velocity (fps) of the water would be most nearly:

(A) 3.5
(B) 3.8
(C) 5.6
(D) 6.8

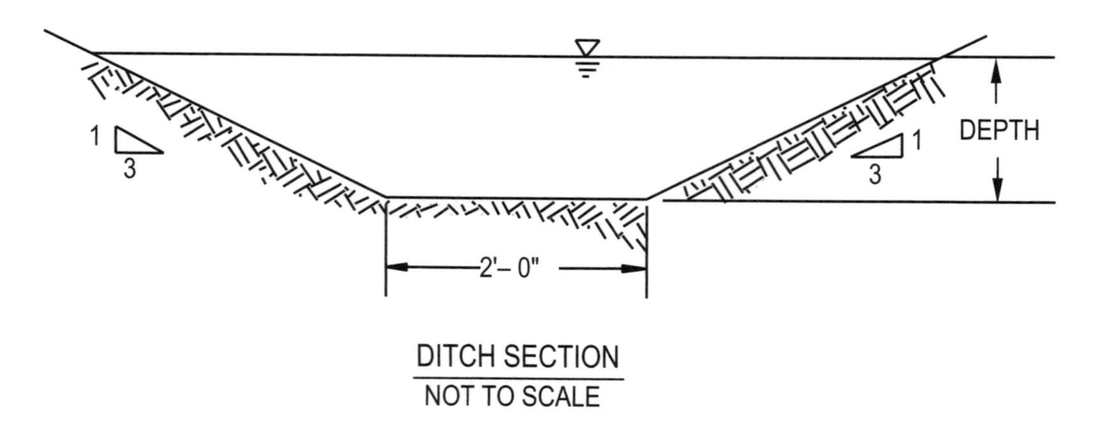

DITCH SECTION
NOT TO SCALE

518. Assume all soils in a drainage basin are in the Soil Conservation Service (SCS) hydrologic soil Group B. Also assume that the vegetative covers are in good condition. The land use is parks and open space. The SCS Runoff Curve Number (CN) for the entire area is most closely approximated by:

(A) 43
(B) 54
(C) 61
(D) 81

GO ON TO THE NEXT PAGE

GO ON TO THE NEXT PAGE

519. A 6.0-MGD water treatment plant is being planned that will use a local river as its water source. The raw water characteristics of the river are shown in **Table 1**.

Tracer studies done on the disinfection tank of a drinking water treatment plant after construction produced the plot shown in the figure below. To satisfy the surface water treatment rule for disinfection of *Giardia* (**Table 2**) on a day when the peak hourly flow rate is 5 MGD, the water temperature is 10°C, and the pH is 7.0, the residual chlorine concentration (mg/L) needed is most nearly:

NOTE: The chlorine dosage must not exceed 1.8 mg/L to minimize THM formation.

(A) 0.60
(B) 0.52
(C) 0.46
(D) 0.42

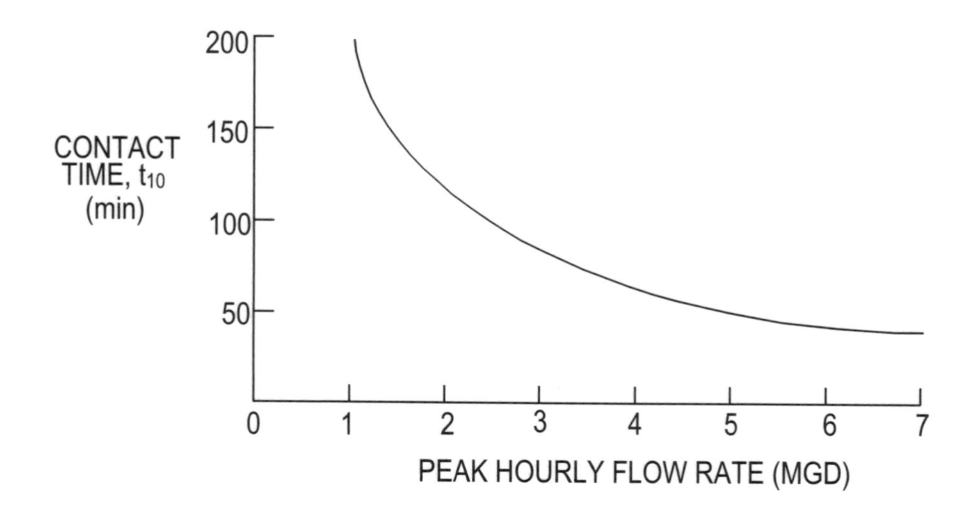

TABLE 1

Parameter	Value
Turbidity	17 NTU
pH	6.8-7.1 S.U.
Alkalinity	150 mg/L as $CaCO_3$
Calcium	51 mg/L as $CaCO_3$
Temperature	10°C to 26°C

TABLE 2

CT values (mg/L•min) for inactivation of *Giardia* cysts by free chlorine at 10°C						
Chlorine Concentration (mg/L)	**pH = 7.0 Log Inactivations**					
	0.5	**1.0**	**1.5**	**2.0**	**2.5**	**3.0**
≤0.4	18	33	33	70	82	105
0.6	19	38	56	75	94	113
0.8	20	40	59	79	99	119
1.0	21	41	62	82	103	123
1.2	21	42	64	85	106	127
1.4	22	44	65	87	109	131
1.6	22	45	67	89	112	134
1.8	23	46	68	91	114	137
2.0	23	46	70	93	116	139
2.2	24	47	71	95	118	142
2.4	24	48	72	96	120	144
2.6	24	49	73	97	122	146
2.8	25	49	74	99	123	148
3.0	25	50	75	100	125	150

Source: U.S. Environmental Protection Agency, *Guidance Manual for Compliance with Filtration and Disinfection Requirements for Public Water Systems Using Surface Water Sources*, Criteria and Standards Division, Office of Drinking Water (U.S.E.P.A. NTIS Publication NO. PB 90-148016), Washington, D.C: U.S. Government Printing Office, October, 1979.

GO ON TO THE NEXT PAGE

520. The results of a well water analysis are given below:

Ca^{++}	51 mg/L
Mg^{++}	12 mg/L
Na^{+}	25 mg/L
$SO_4^{=}$	65 mg/L
Cl^{-}	25 mg/L
F^{-}	0.4 mg/L
NO_3^{-}	14 mg/L as NO_3^{-}–N
pH	7.8
H_2S	3.4 mg/L as S
Alkalinity	84 mg/L as $CaCO_3$
Total coliforms	MPN 2.2 org/100 ml
Turbidity	6.2 NTU
Chlorine demand	9.1 mg/L
TDS	332 mg/L
Temperature	25°C

The contaminants that exceed the primary or secondary maximum contaminant levels are:

(A) nitrate, coliform, and turbidity
(B) fluoride, coliform, and turbidity
(C) TDS, turbidity, and fluoride
(D) hydrogen sulfide, TDS, and coliform

GEOTECHNICAL

AFTERNOON SAMPLE QUESTIONS

This book contains 20 geotechnical depth questions, half the number on the actual exam.

In some cases the same question appears in more than one of the depth modules because there is crossover of knowledge between the depth areas of civil engineering. This use of the same question on different depth modules could occur on the examination.

GEOTECHNICAL AFTERNOON SAMPLE QUESTIONS

Questions 501–502: Laboratory testing was performed on two soil samples (Samples A and B) and the data is summarized in the table.

Sieve Analysis Data and Index Properties of Sample A and Sample B		
Sieve #	**Sample A % passing**	**Sample B % passing**
3 inches	100	
1 1/2 inches	98	
3/4 inch	96	
#4	77	100
#10	—	96
#20	55	94
#40	—	73
#100	30	—
#200	18	55
Liquid limit	32	52
Plastic limit	25	32

501. According to the Unified Soil Classification System, the classification of Sample A is most nearly:

(A) SW
(B) SP
(C) SM
(D) SC

502. Sample B can be classified by the AASHTO system as:

(A) A-5
(B) A-6
(C) A-7-5
(D) A-7-6

GEOTECHNICAL AFTERNOON SAMPLE QUESTIONS

503. A cylindrical specimen is tested in the laboratory. The following properties were obtained:

Sample diameter	3 inches
Sample length	6 inches
Weight before drying in oven	2.95 lb
Weight after drying in oven	2.54 lb
Oven temperature	110°C
Drying time	24 hours
Specific gravity	2.65

The degree of saturation of Sample A is most nearly:

(A) 60%
(B) 70%
(C) 80%
(D) 90%

504. A saturated cohesionless soil was tested in a triaxial apparatus.

Data at Failure:	
Pore water pressure	10.0 psi
Vertical total stress	33.5 psi
Horizontal total stress	16.4 psi

Under drained conditions, the effective friction angle (degrees) is most nearly:

(A) 25
(B) 30
(C) 35
(D) 40

GO ON TO THE NEXT PAGE

GO ON TO THE NEXT PAGE

Questions 505–506: A reinforced concrete dam having a sheet-pile cutoff wall is to be used to retain 12-foot-high water. The dam is 120 feet long and the dimensions and the flow net are shown in the figure below. The granular soil has the following characteristic:

Coefficient of permeability 0.003 fps (isotropic soil)

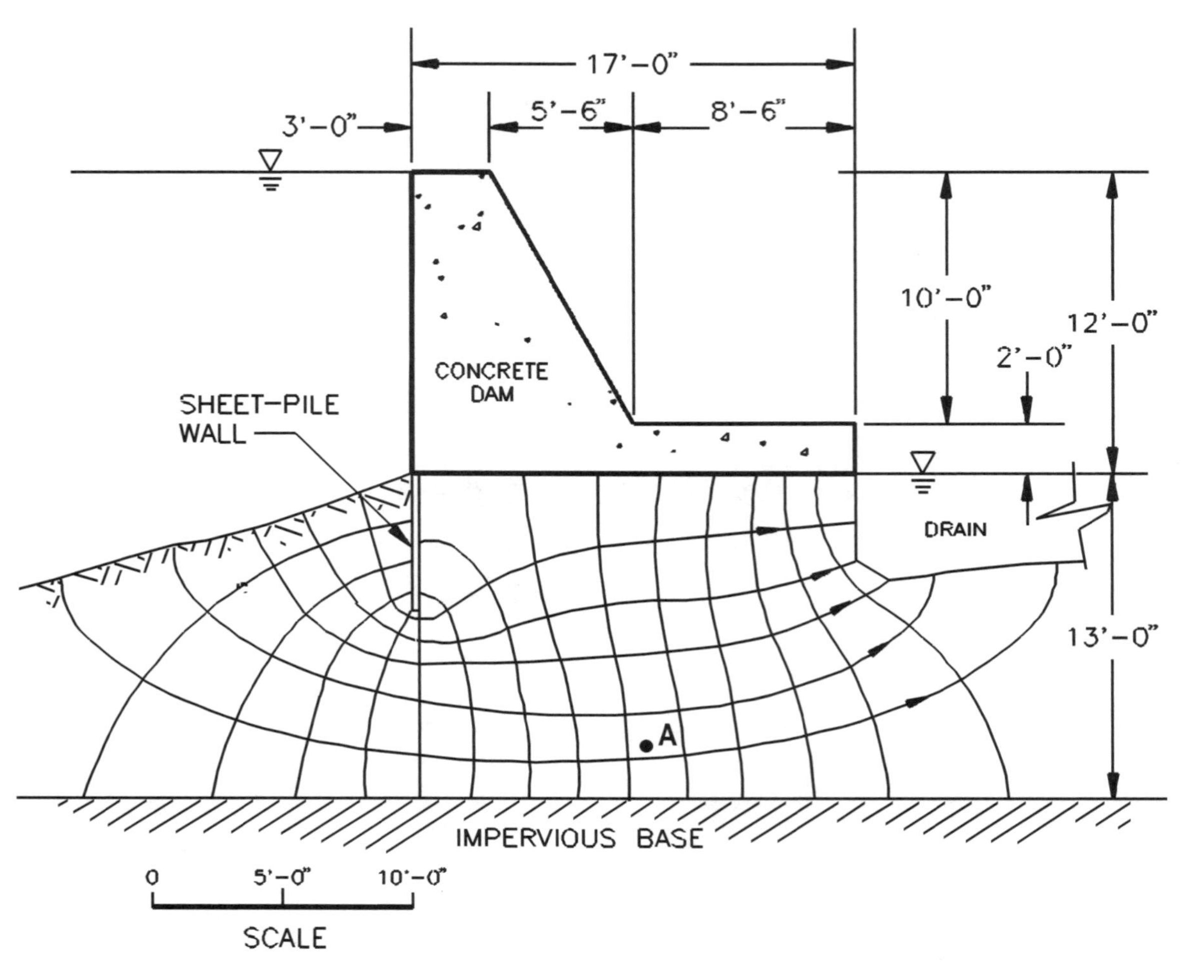

CONCRETE DAM SECTION

505. The quantity of seepage loss under the dam (cfs) is most nearly:

(A) 0.014
(B) 1.4
(C) 1.7
(D) 17.0

506. Point A is 2.0 feet above the impervious base. The pressure head at Point A (feet) is most nearly:

(A) 4.4
(B) 7.6
(C) 11.0
(D) 15.4

GO ON TO THE NEXT PAGE

Questions 507–508: A mat foundation will be constructed at ground surface. The subsoil profile is shown in the figure below.

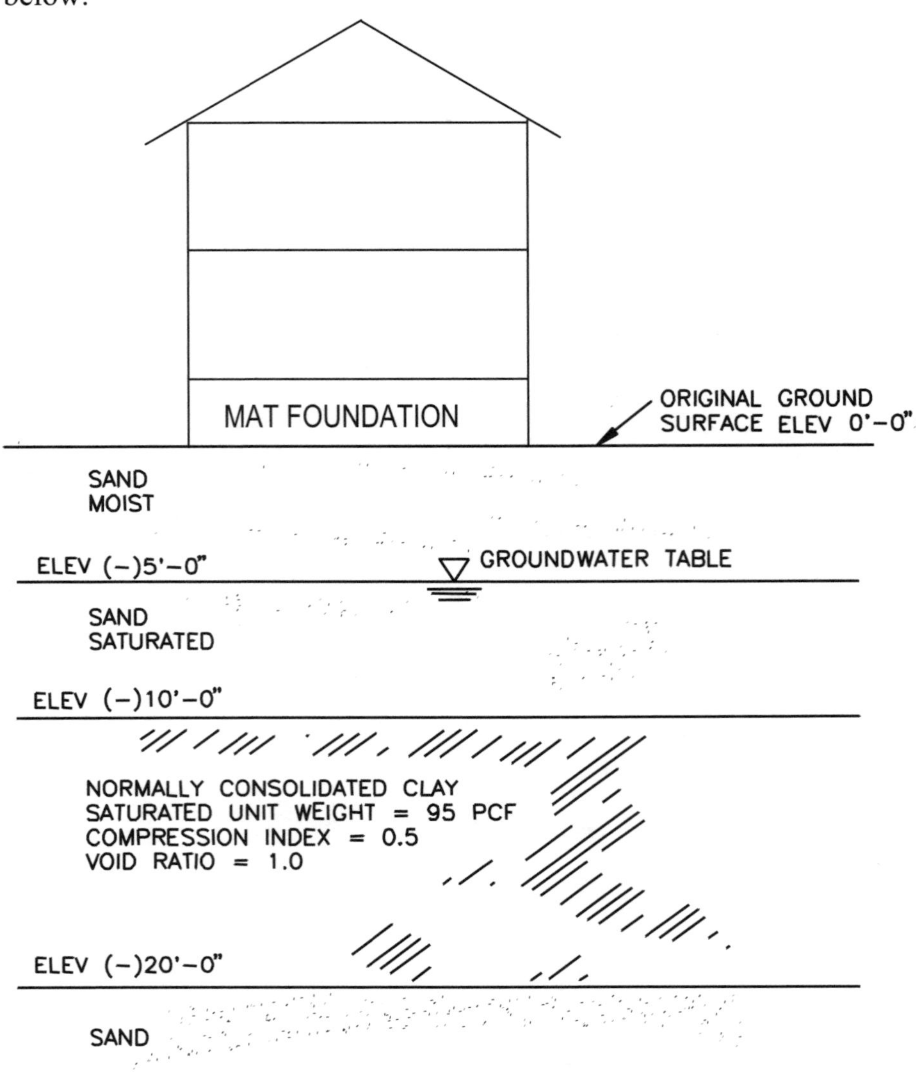

GO ON TO THE NEXT PAGE

507. Assuming that the initial effective overburden pressure is 1,000 psf and the stress increase due to the mat foundation is 200 psf at the middle of the clay layer, the primary consolidation settlement (inches) of the clay layer is most nearly:

(A) 0.2
(B) 1.0
(C) 2.4
(D) 3.6

508. Assuming only vertical drainage and that the coefficient of consolidation is 1.6×10^{-4} in^2/sec, the time (months) required to achieve 50% primary consolidation is most nearly:

(A) 2
(B) 8
(C) 12
(D) 60

GO ON TO THE NEXT PAGE

Questions 509–510: A 13-foot, 12-inch-diameter round timber pile is to be driven into the soil as shown in the figure below.

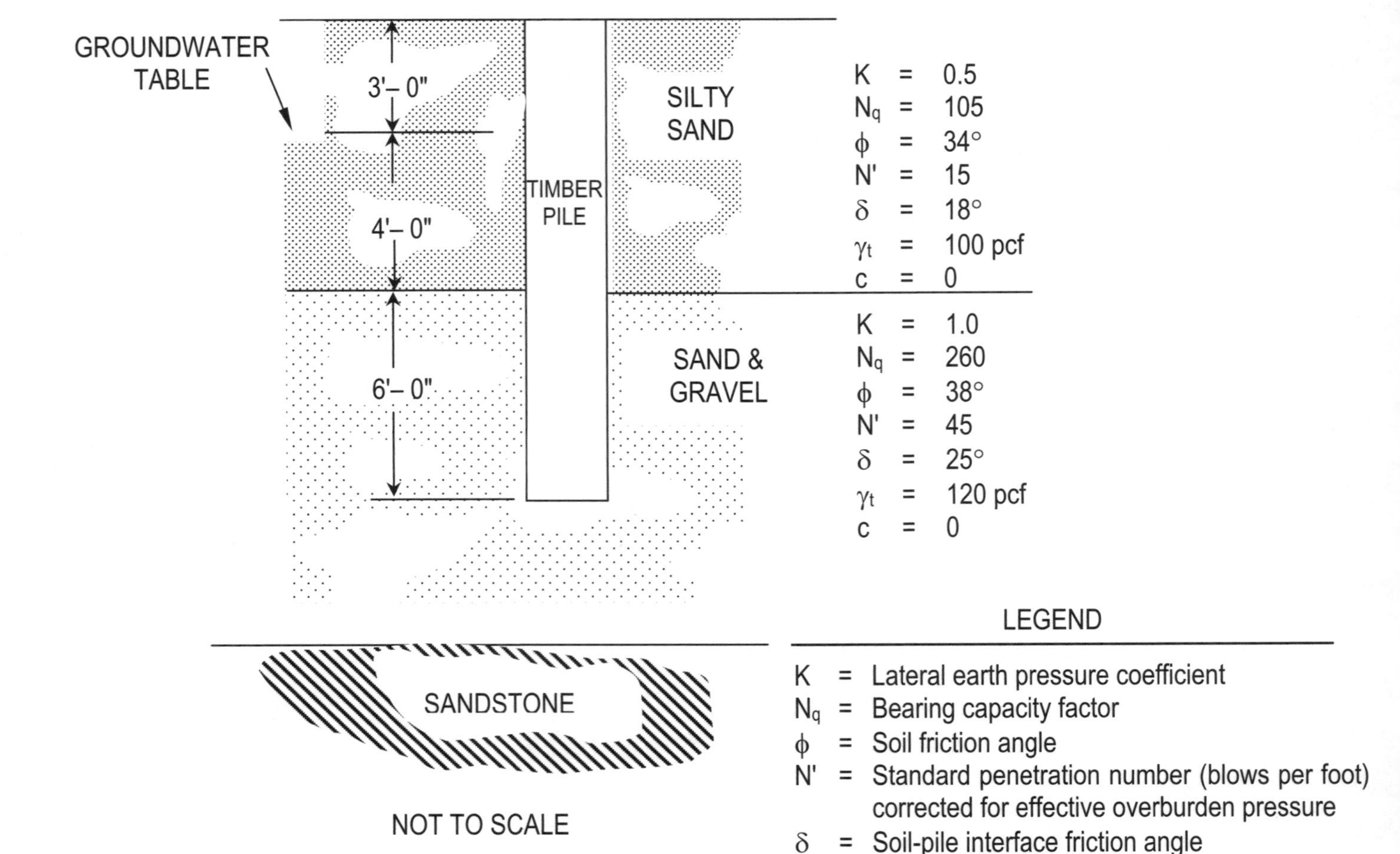

GO ON TO THE NEXT PAGE

509. From the information given, the ultimate end-bearing capacity (tons), using the Terzaghi method, is most nearly:

(A) 76
(B) 82
(C) 146
(D) 164

510. By using the Mohr-Coulomb equation, the ultimate side-friction capacity (tons) is most nearly:

(A) 2.9
(B) 3.3
(C) 3.9
(D) 5.8

GO ON TO THE NEXT PAGE

Questions 511–512: A 12-foot-high road embankment will be built using the reinforced earth method to resist lateral earth pressure as shown in the figure below.

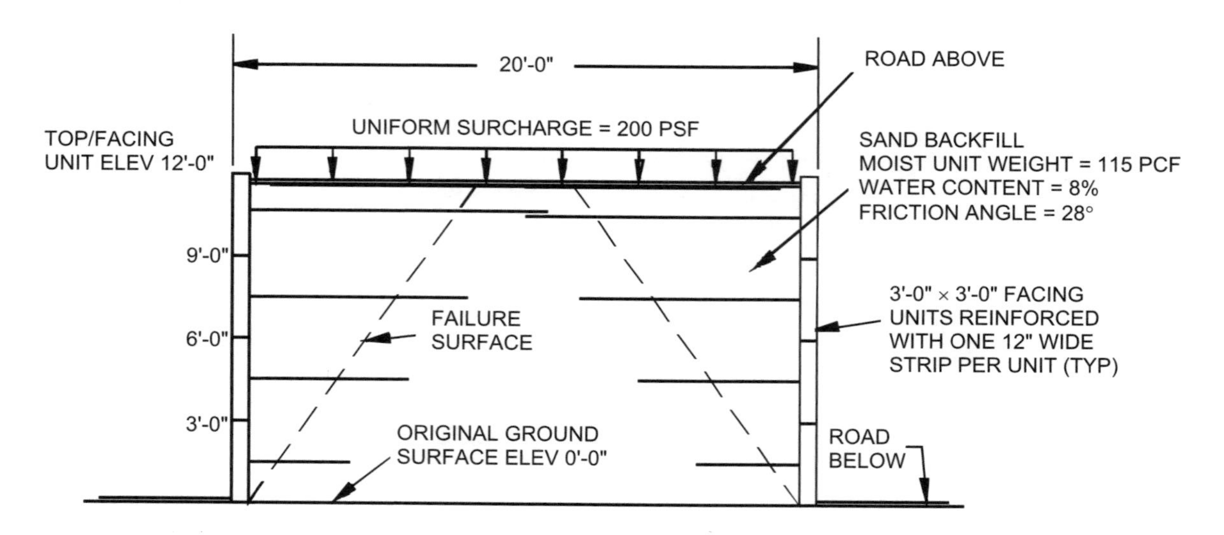

511. Assuming the active lateral earth pressure coefficient is 0.3, the resultant force (lb) due to the active lateral earth pressure force acting on the facing unit located from Elev. 0'-0" to 3'-0" is most nearly:

(A) 1,300
(B) 3,200
(C) 3,400
(D) 3,800

512. Assume the active lateral earth coefficient of the sand backfill is 0.3, the friction angle between the reinforcing strip and the sand is 22°, and the required safety factor against slipping of the reinforcing strip is 3. The distance (feet) that the bottom reinforcing strip needs to be extended beyond the failure surface is most nearly:

(A) 6
(B) 8
(C) 10
(D) 12

GO ON TO THE NEXT PAGE

513. The figure below summarizes the findings of a subsurface investigation program conducted at a potential site for a proposed bridge crossing. The liquefaction potential of the site is to be evaluated.

Assume that the effective overburden pressure at mid-depth of Layer No. 3 is 1.10 tsf, the earthquake-induced average shear stress is 450 psf, and the cyclic stress ratio is 0.29. The factor of safety against liquefaction is most nearly:

(A) 0.7
(B) 1.2
(C) 1.3
(D) 1.4

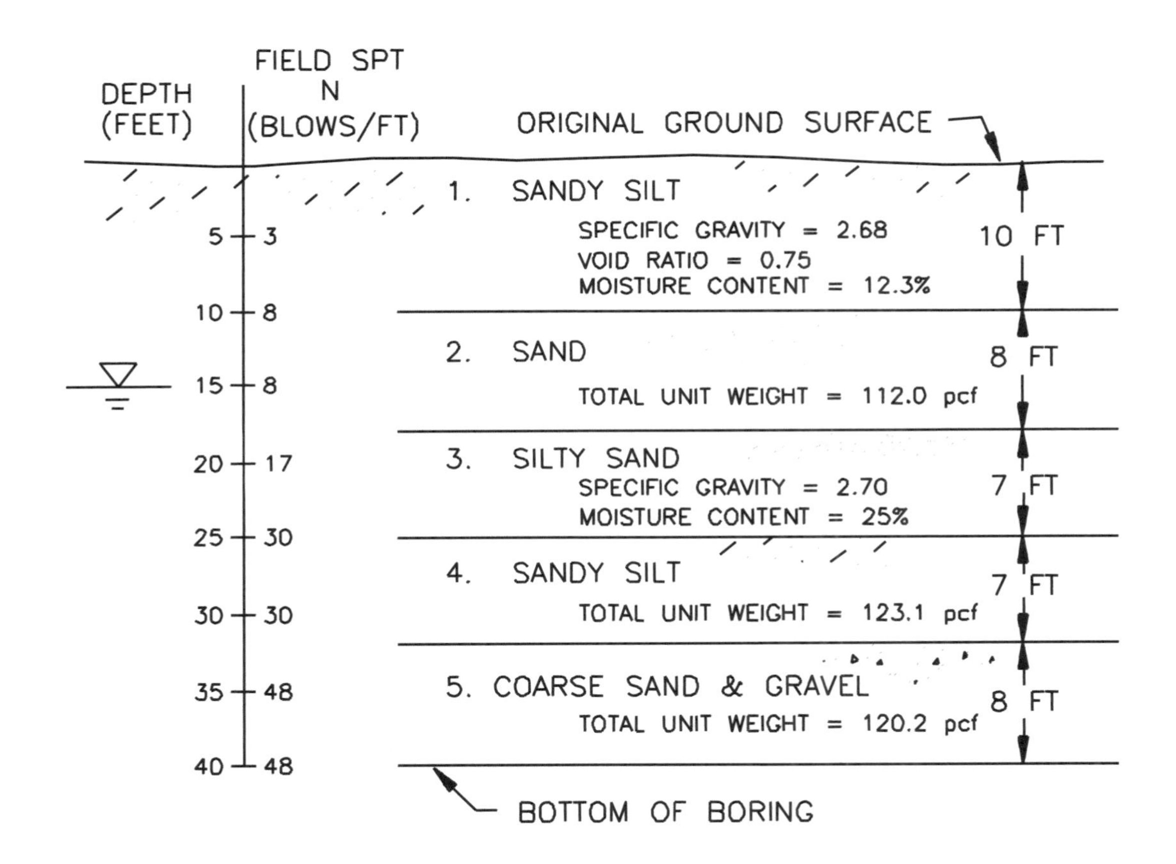

SUBSURFACE INVESTIGATION SUMMARY
NOT TO SCALE

Questions 514–515: The geological conditions shown in the figure below were encountered during the initial reconnaissance of an industrial property intended for landfill development. Testing of the soil materials encountered revealed the following characteristics.

Fractured clay till - El. 870–890 MSL (all elevations are in feet MSL):

In-situ horizontal permeability	1×10^{-5} cm/sec
Remolded permeability at 90% of maximum dry density and optimum moisture	9×10^{-8} cm/sec
Optimum moisture content at maximum dry density	8%
In-situ moisture content	11%
Plasticity Index	15
USCS classification	SC

Unfractured Lacustrine Clay - El 840–870 MSL

In-situ horizontal permeability	1×10^{-8} cm/sec
USCS classification	CL

Well A is screened from 890–895 MSL. Well B is screened from 820–825 MSL. Over a 3-month period, the static water level has been measured to average 895 MSL in Well A, and to average 898 MSL in Well B. A stream within 300 feet of the site has a measured normal high-water elevation of 875 MSL. The industrial waste to be disposed in the proposed landfill is non-hazardous for purposes of RCRA classification using TCLP testing, but does exhibit significant potential to leach metal contaminants and create groundwater contamination if not properly handled.

Assume well casings have been properly grouted into the formations.

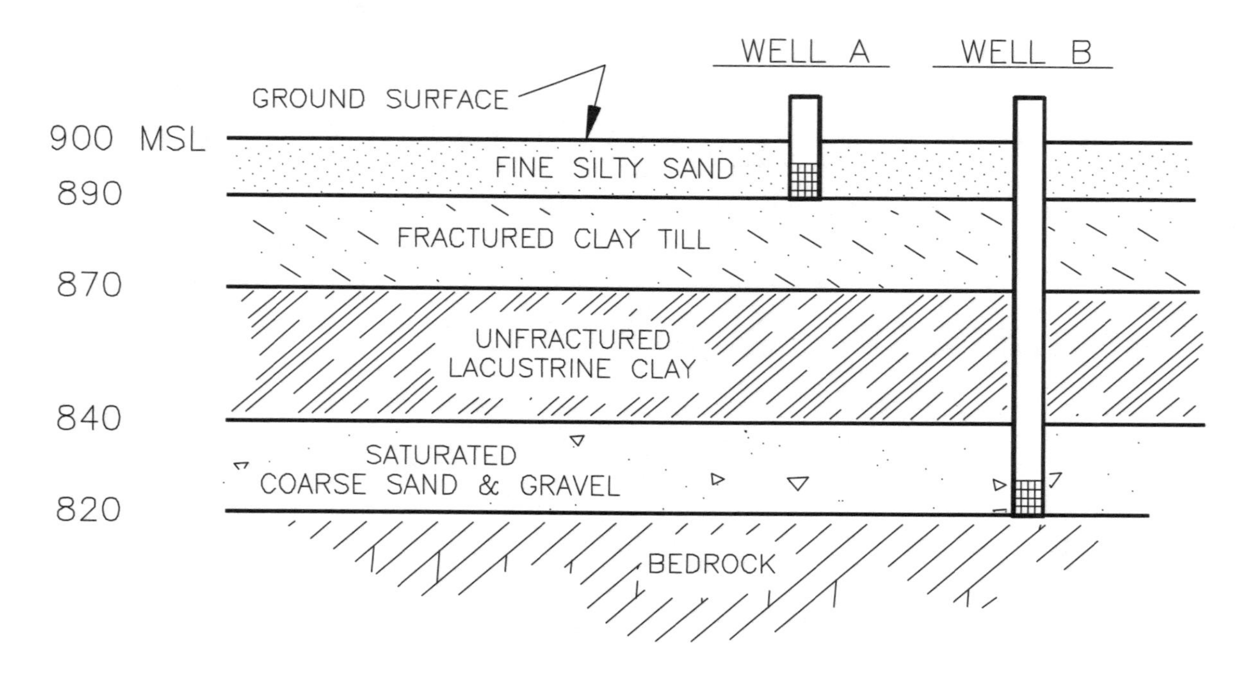

NOT TO SCALE

514. The formation between El. 820–840 MSL would be most accurately characterized as:

(A) a flowing artesian aquifer
(B) a deep water-table aquifer
(C) an interbedded lacustrine aquiclude
(D) a confined aquifer

515. The most economical way to control lateral groundwater migration through the upper silty sand on the site would be to:

(A) eliminate the shallow groundwater in the vicinity of the site by using gravity dewatering

(B) increase the isolation to the groundwater by building up the grade on the site to elevation 905 MSL before beginning construction of the landfill

(C) use slurry walls around the perimeter of the site to isolate the landfill from the shallow groundwater

(D) install a leak-proof liner to isolate the waste material from the environment

GO ON TO THE NEXT PAGE

Questions 516–517: Non-air-entrained concrete, designed in accordance with the Portland Cement Association, is required for a slab foundation. Two trial batch mixes have been prepared and tested, and the data is summarized in the table below. The aggregates used in the trial batch mixes were wet, with the fine aggregate having a moisture content of 6.5%, and the coarse aggregate having a moisture content of 2.5%. Laboratory tests performed on the aggregates are summarized in the laboratory test data shown on this page.

Trial Mix Data						
Trial Batch Mix No.	**Weight of Wet Coarse Aggregate (lb)**	**Weight of Wet Fine Aggregate (lb)**	**Weight of Added Cement (lb)**	**Weight of Added Water (lb)**	**Volume of Consolidated Concrete (ft^3)**	**f'_c (psi)**
1	130.0	100.0	50.0	30.0	2.27	4,500
2	129.0	101.0	55.0	30.3	2.30	5,000

Laboratory Test Data:

1) Saturated Surface Dry Moisture Content:
   Coarse Aggregate = 0.50%
   Fine Aggregate = 1.5%

2) Saturated Surface Dry Specific Gravity:
   Coarse Aggregate = 2.65
   Fine Aggregate = 2.70

Consider the concrete in the final consolidated state.

GO ON TO THE NEXT PAGE

516. For Trial Mix No. 1, the unit weight (pcf) is most nearly:

(A) 132
(B) 137
(C) 140
(D) 145

517. For Trial Mix No. 2, the percent volume of aggregate is most nearly:

(A) 28%
(B) 38%
(C) 48%
(D) 58%

Questions 518–519: A caisson wall will be built to stabilize an active landslide as shown in the figure on the opposite page. The caissons will be 3'-0" in diameter and spaced 5'-0" on-center. After excavation of the caisson hole and placement of the steel section, the hole will be backfilled to the top of rock with high-strength concrete. The remaining portion of the hole will be backfilled up to the ground surface with a low-strength sand-cement slurry, the strength of which is to be neglected. The caisson wall will be installed with tiebacks.

Subsurface exploration revealed that the ground surface, the slide plane, and the solid-rock surface all slope at a ratio of 3:1 (horizontal:vertical). The slide plane is 18'-0" below the ground surface, and the solid rock surface is 20'-0" below the ground surface.

Assumptions:

- Landslide will subject the caisson wall to a uniform pressure, p, that is inclined to the horizontal.
- Soil arching will occur between the caissons to transfer the landslide pressure to the caissons.
- Vertical component forces from the landslide and the tieback are concentric on the caissons.
- Point of fixity for the caisson design is at 1'-0" below the top of rock.
- Soil provides complete lateral support for the steel section; i.e., zero unbraced length.
- Neglect the weight of the caissons.

Also assume a safe design landslide pressure of $p = 1.0$ kip/ft^2 (unfactored) with a simultaneous tieback tension of 10 kips (unfactored). Each caisson has a steel tieback attached to the steel section at the ground surface (see figure on the opposite page) and anchored into a grouted 6-inch-diameter drilled hole in the solid rock. Based on previous load tests into this same rock mass, the ultimate frictional resistance at pullout is 10 psi. The minimum factor of safety is 1.5 against pullout. The tieback is to be tensioned after installation.

518. The maximum unfactored bending moment (ft-kip) in the caisson is most nearly:

(A) 1,220
(B) 1,030
(C) 840
(D) 760

519. Assuming the safe tension force is 10 kips for the tieback, the minimum bonded length (feet) of the tieback anchorage into solid rock is most nearly:

(A) 10
(B) 7
(C) 5
(D) 3

GO ON TO THE NEXT PAGE

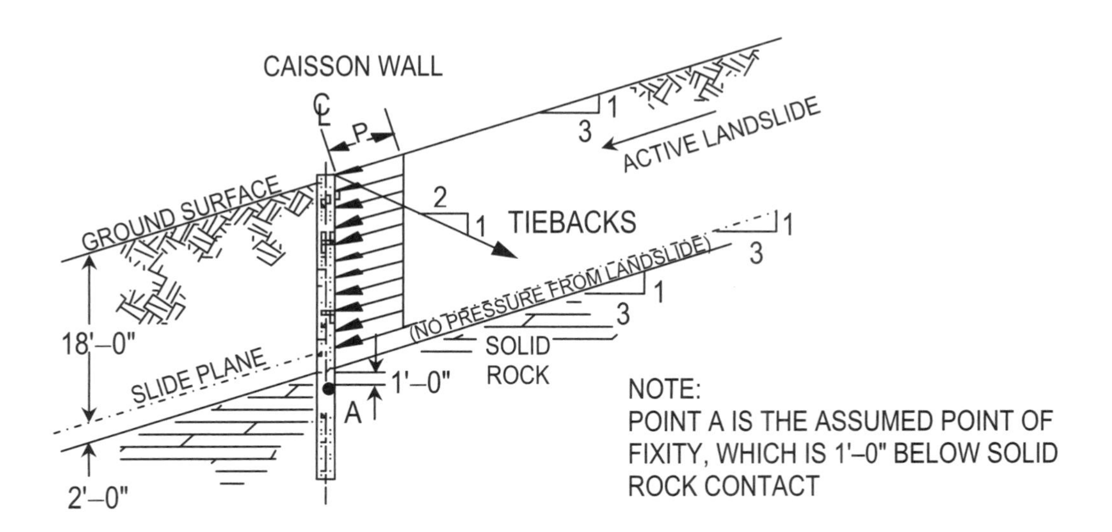

CAISSON WALL SECTION

NOT TO SCALE

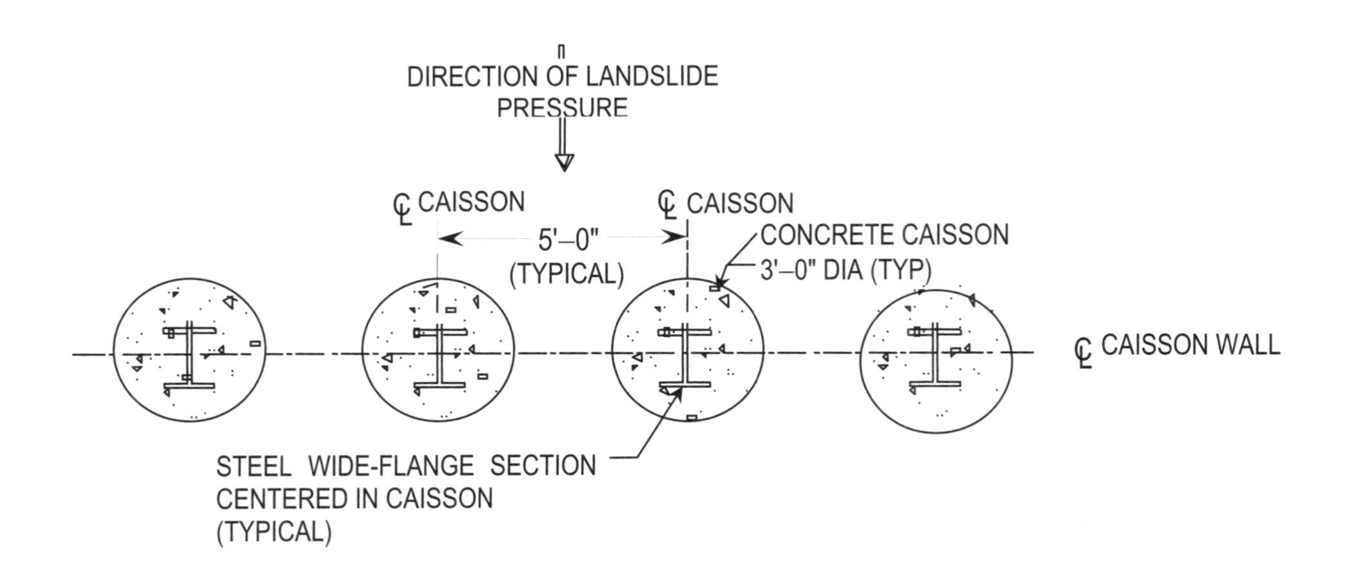

TYPICAL CAISSON WALL PLAN

NOT TO SCALE

GO ON TO THE NEXT PAGE

Geotechnical Question 520 also appears as Structural Question 519.

520. A segment of interstate highway requires the construction of an embankment of 500,000 yd^3. The embankment fill is to be compacted to a minimum of 90% of Modified Proctor maximum dry density.

A source of suitable borrow has been located for construction of the embankment. Assume that there is no soil loss in transporting the soil from the borrow pit to the embankment.

Use the following data:

Dry unit weight of soil in borrow pit	113.0 pcf
Moisture content in borrow pit	16.0%
Specific gravity of the soil particles	2.65
Modified Proctor optimum moisture content	13.0%
Modified Proctor maximum dry density	120.0 pcf

Assuming each truck holds 5.0 yd^3 and the void ratio of the soil is 1.30 during transport, the minimum number of truckloads of soil from the borrow pit that is required to construct the embankment is most nearly:

(A) 100,000
(B) 150,000
(C) 200,000
(D) 250,000

STRUCTURAL

AFTERNOON SAMPLE QUESTIONS

This book contains 20 structural depth questions, half the number on the actual exam.

In some cases the same question appears in more than one of the depth modules because there is crossover of knowledge between the depth areas of civil engineering. This use of the same question on different depth modules could occur on the examination.

For **Questions 501 and 502**, refer to the figure below. The concrete unit weight is 150 pcf. The wall panels have a 3'-0" parapet and a total height of 26'-0". The roof dead load including trusses and glulams is 15 psf. The glazing weight is 10 psf. The building is located in Seismic Zone 3 ($A_a = A_v = 0.30$), and the soil profile is not known in sufficient detail. This is a standard occupancy structure.

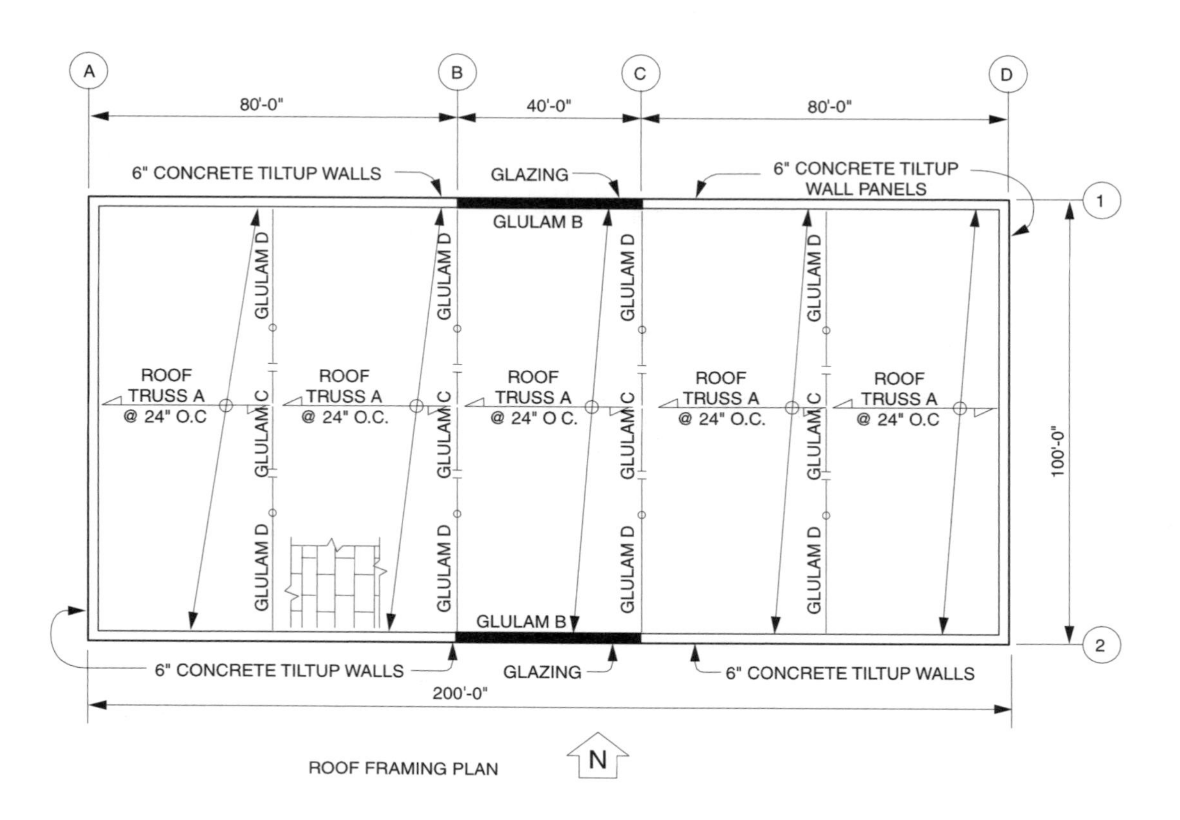

ROOF FRAMING PLAN

STRUCTURAL AFTERNOON SAMPLE QUESTIONS

501. The total seismic dead load W (kips) at the roof level is most nearly:

(A) 440
(B) 510
(C) 660
(D) 880

502. The base shear as a fraction of the total seismic dead load W (C_s in the NBC or SBC) is most nearly:

	UBC	NBC or SBC
(A)	0.16	0.14
(B)	0.20	0.17
(C)	0.57	0.27
(D)	0.88	0.45

GO ON TO THE NEXT PAGE

503. The figure below illustrates a cross-section of a deck slab of a steel girder bridge that is exposed to deicing salt. For an HS20 live load, steel reinforcement $F_y = 60{,}000$ psi and $f'_c = 4{,}000$ psi. AASHTO *Standard Specifications for Highway Bridges*, 16th edition (1996, with interims through 1998), applies.

For an effective span length of 9'-6", the live load moment (ft-kips), excluding impact, is most nearly:

(A) 3.5
(B) 4.6
(C) 5.8
(D) 9.2

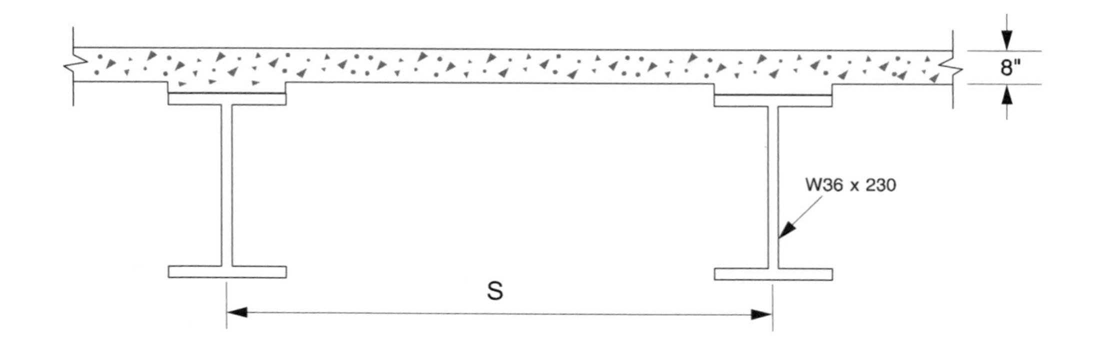

CROSS SECTION

GO ON TO THE NEXT PAGE

The following information applies to **Questions 504–505**:

Codes:
ACI 530-95, *Building Code Requirements for Masonry Structures*

Materials:
Hollow concrete masonry units f'_m = 1,500 psi with Type S mortar. Cells with reinf. grouted
Steel reinforcement ASTM A615 Grade 60

Loads:
Roof dead load = 15 psf
Non reducible roof snow load = 40 psf
Average wall dead load = 54 psf
Design wind (pressure or suction) = 20 psf
Seismic forces do not govern.

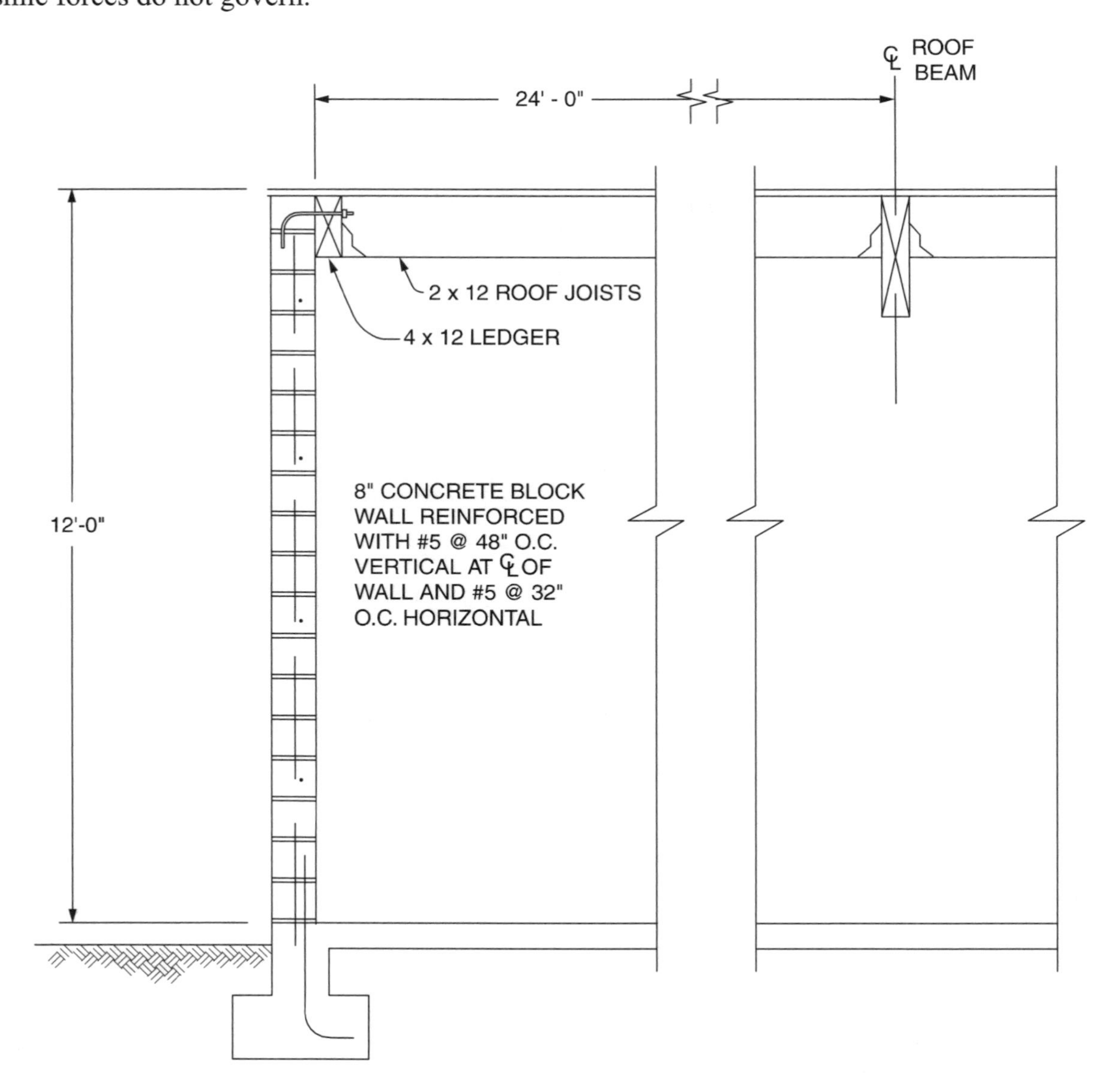

504. The maximum design moment (lb-ft/ft) for the masonry wall is most nearly:

(A) 160
(B) 360
(C) 560
(D) 760

505. The maximum allowable moment (lb-ft/ft) on the masonry wall based on the maximum allowable masonry flexural stress is most nearly:

(A) 665
(B) 885
(C) 1,465
(D) 2,335

506. The figure on the opposite page shows the elevation and plan view of a single-story building with precast, prestressed, double-tee wall panels and double-tee roof elements supported on the wall panels on the outside and on an inverted tee beam on the inside. Section 1-1 shows the cross section of the roof elements.

Assume wind loading only and that the wind is in the north-south direction. Design wind pressure, including shape factors = 25 psf.

If the wind load w is 400 plf, the maximum roof diaphragm service load chord force (kips) is most nearly:

(A) 3.8
(B) 15.4
(C) 30.7
(D) 48.0

The following figure applies to **Question 506**.

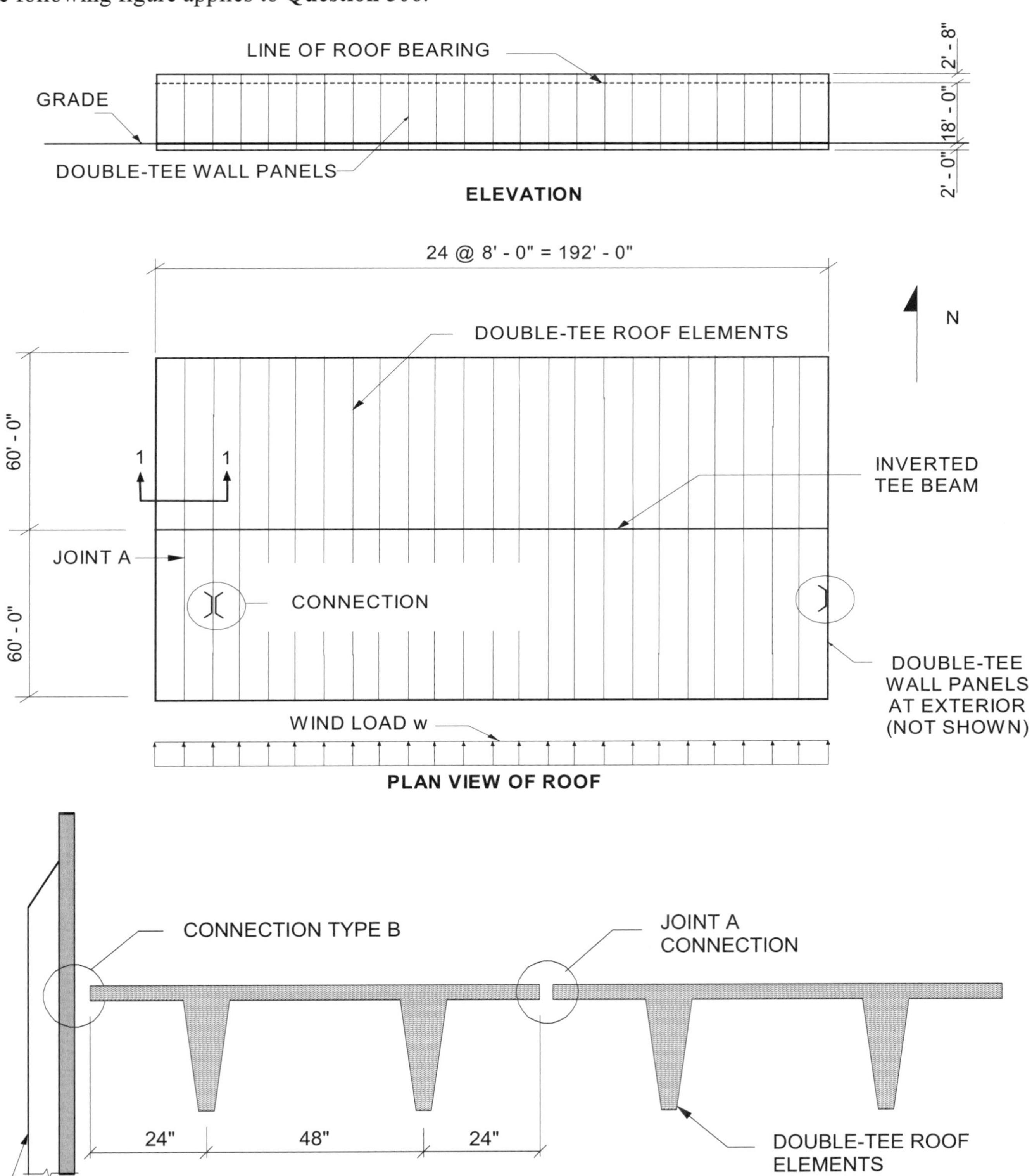

507. Assume:

No earthquake forces
No ice build-up
Pipe shape factor for wind = 0.67

Sign (reader board) dead load = 20 psf
(includes support arm)

Wind force = 36 psf

Support pipe column:

Diameter = 12 inches O.D.
Weight = 46.6 plf
Area = 13.7 in^2
Section modulus = 38.6 in^3
Moment of inertia
$I_{x\text{-}x} = I_{y\text{-}y} = 231\ in^4$

Any reference to "sign" in the question refers to complete sign/pipe column assembly.

For the dimensions given below, the maximum shearing stress (ksi) in the pipe column when the sign is subjected to the wind load stated previously is most nearly:

x = 4'-0"
y = 8'-0"

X = 22'-0"
Y = 15'-0"

	ASD	LRFD
(A)	2.90	3.70
(B)	4.10	5.30
(C)	5.50	7.20
(D)	8.00	10.50

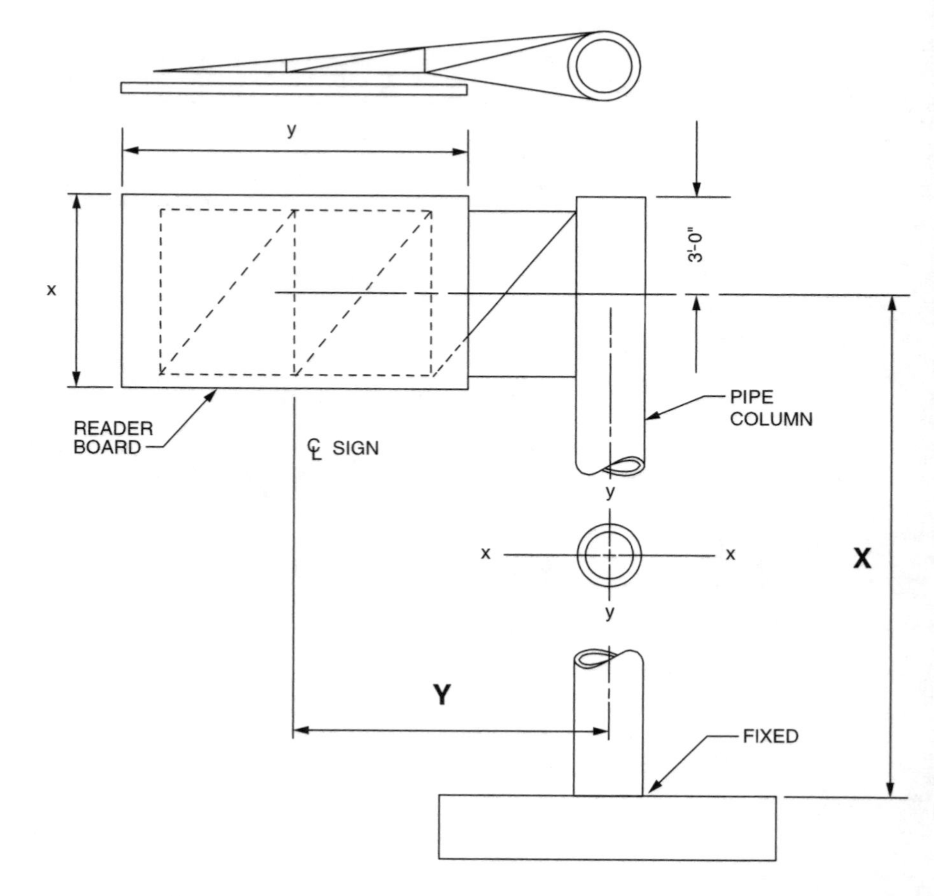

GO ON TO THE NEXT PAGE

GO ON TO THE NEXT PAGE

The following information applies to **Question 508**:

The figure below shows a typical steel rigid frame of a warehouse. Point B of Column AB is braced against movement perpendicular to the plane of the frame. Point A is attached to a simple spread footing by anchor bolts and can be considered hinged about both axes. The flanges of Column AB are laterally supported only at Points A and B, since full-height loading doors are located on both sides of Column AB. Point B is subject to sidesway in the plane of the frame.

A W18 × 40 section has already been selected for Girder BC. W14 × 53 has been selected for the column section.

Column AB is subjected to the following axial compression forces and bending moments at Point B about the major axis:

Load	Compressive Axial Force, kips	Moment, ft-kips
Dead	7.2	40
Live	12.6	60
Wind	4.0	70

The AISC-ASD *Manual of Steel Construction*, 9th edition, or AISC-LRFD *Load and Resistance Factor Design*, 2nd edition, applies.

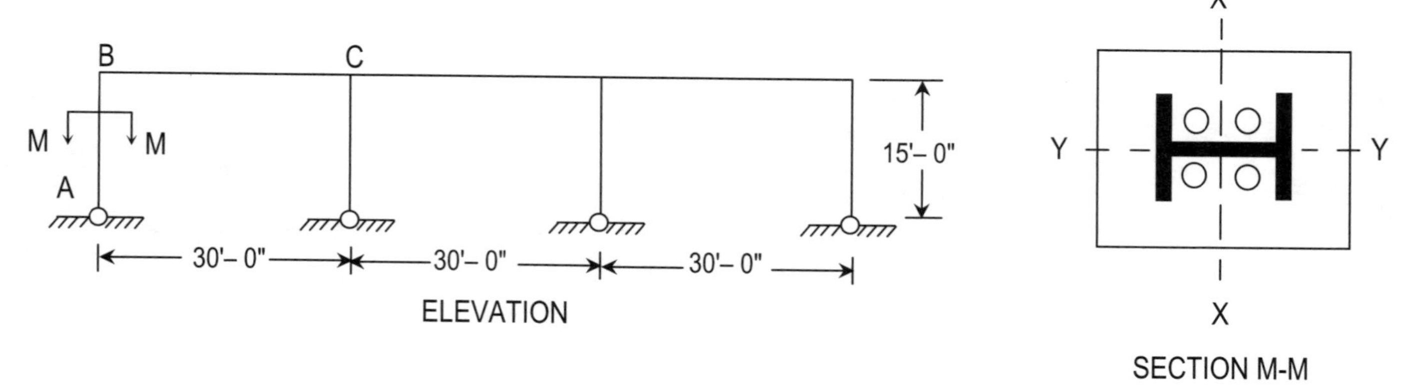

You may select **EITHER** the ASD **OR** the LRFD option.

508. ASD option:

For $f_b = 20.0$ ksi and $F_b = 24.0$ ksi (including any permitted stress increase), the critical combined stress ratio for Column AB under D + L + W load combination, with $K_x = 2.0$, is most nearly:

(A) 0.917
(B) 0.944
(C) 0.986
(D) 1.042

LRFD option:

For $\phi_b M_{nx} = 235$ ft-kips and $M_{ux} = 180$ ft-kips (including any permitted stress increase), the critical combined stress ratio for Column AB under D + L + W load combination, with $K_x = 2.0$, is most nearly:

(A) 0.819
(B) 0.849
(C) 1.018
(D) 1.103

GO ON TO THE NEXT PAGE

509. When a concrete slab is placed on a hot windy day, it is **NOT** permissible to:

(A) add field water as needed to obtain the desired consistency and workability

(B) keep mix water cool and aggregate moist by shading and sprinkling

(C) spray or protect the concrete surfaces with wet burlap to retard hardening

(D) moisten the forms and the reinforcement prior to placement of concrete to minimize evaporation

510. The figure below illustrates a cross-section of a composite two-lane, 85-foot span, simply-supported highway bridge. AASHTO *Standard Specifications for Highway Bridges*, 16th edition (1996, with interims through 1998) applies.

For an HS20-44 truck, the live load plus impact moment (ft-kips) of an interior girder is most nearly:

(A) 2,400
(B) 1,200
(C) 780
(D) 500

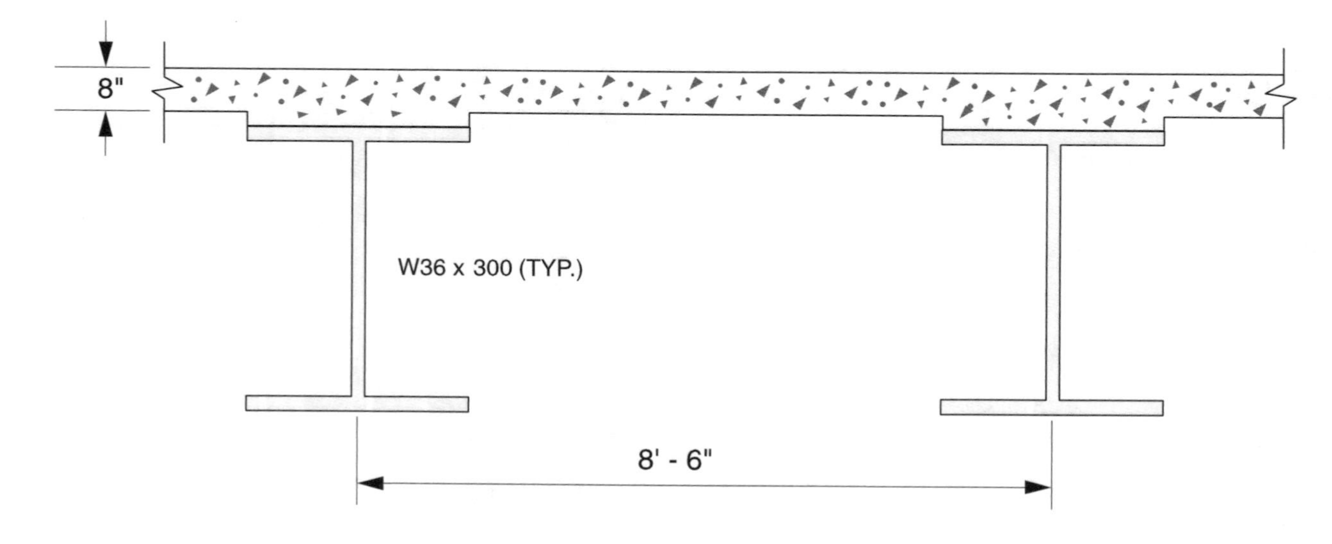

GO ON TO THE NEXT PAGE

511. A steel canopy, supported by a single round column with four anchor bolts, is shown in the figure. Ignore unbalanced vertical loads and weight of canopy. The minimum diameter (inches) of an A307 anchor bolt is most nearly:

(A) 7/8
(B) 1
(C) 1 1/8
(D) 1 1/4

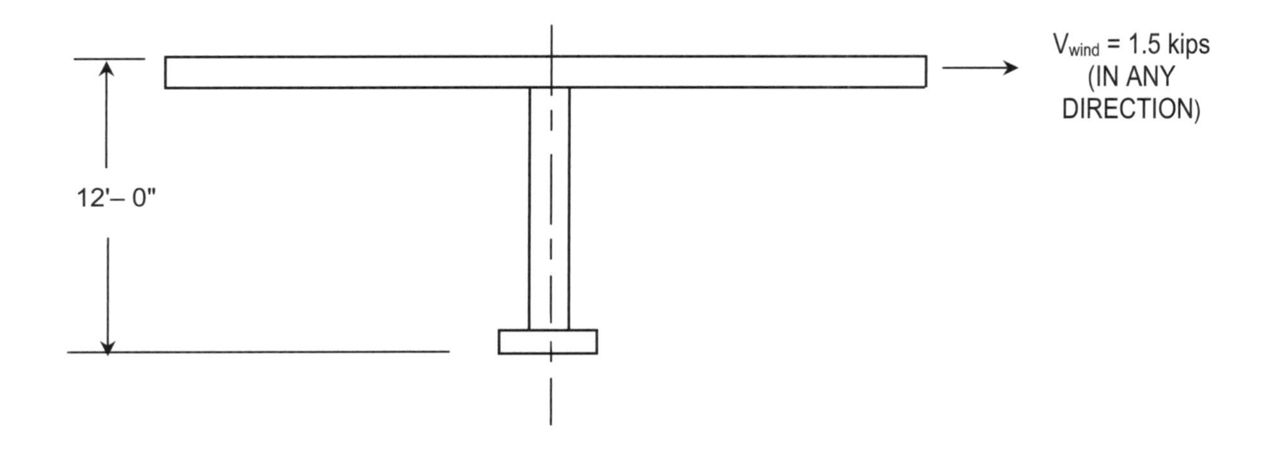

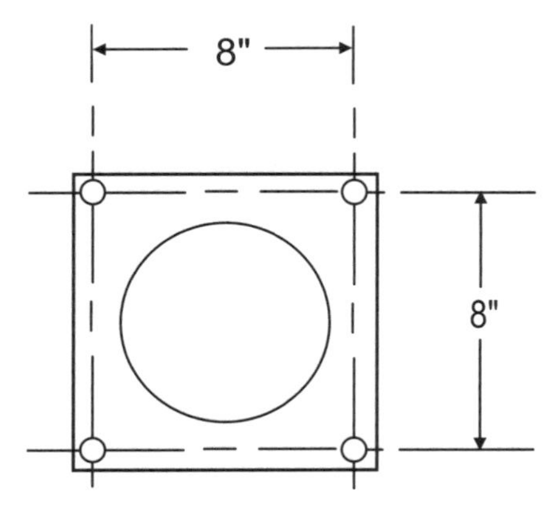

BASE PLATE PLAN

512. The figures on this page show the plan and elevation of a square footing supporting a column. The following data are available:

Service loads per the figure:

Net allowable soil bearing pressure = 4,000 psf (silty-sandy soil)

f'_c, normal weight concrete = 3,000 psi

Reinforcing steel = Grade 60, ASTM A-615

The service live load moment acts in either perpendicular direction. However, it acts in one direction at a time. The strength design provisions of the ACI *Building Code Requirements for Reinforced Concrete*, ACI 318-95, apply. Neglect the weight of the soil and footing in the calculations.

The minimum dimension required (feet) for the footing is most nearly:

(A) 3.7
(B) 5.5
(C) 5.9
(D) 7.0

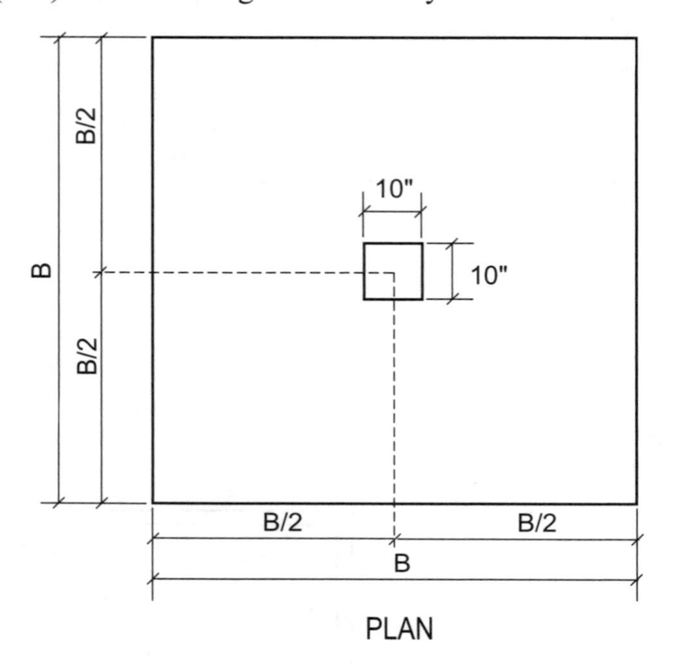

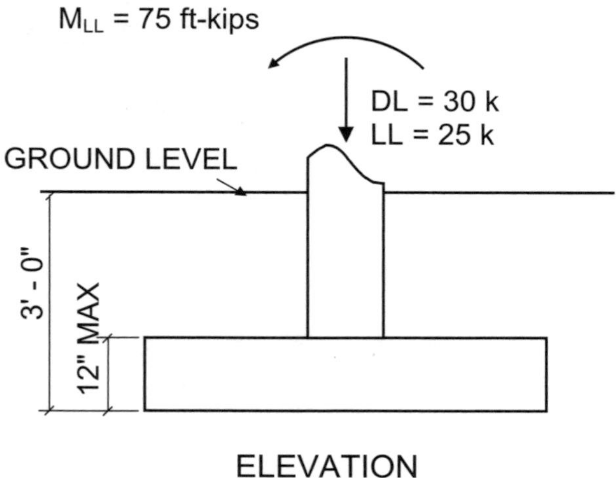

GO ON TO THE NEXT PAGE

513. The structural glued, laminated timber member shown in the figure below is subjected to a total uniform lateral load of 200 plf, including the member weight, and an axial compression force of 15,000 lb. The member is simply supported with the ends held in position by other framing members; no other lateral support is to be relied upon. The expected service moisture content is less than 16%, and the temperature is less than 100°F. The laminated timber combination symbol is 24F-V2 with Hem-Fir laminations.

The *National Design Specifications for Wood Construction*, 1991 edition, and the following data apply:

E_{xx} = 1,500,000 psi

E_{yy} = 1,400,000 psi

Under normal load duration, for a trial-size member dimension of 10 3/4 inches wide by 18 inches deep, the critical buckling design value (psi) is most nearly:

(A) 520
(B) 400
(C) 375
(D) 100

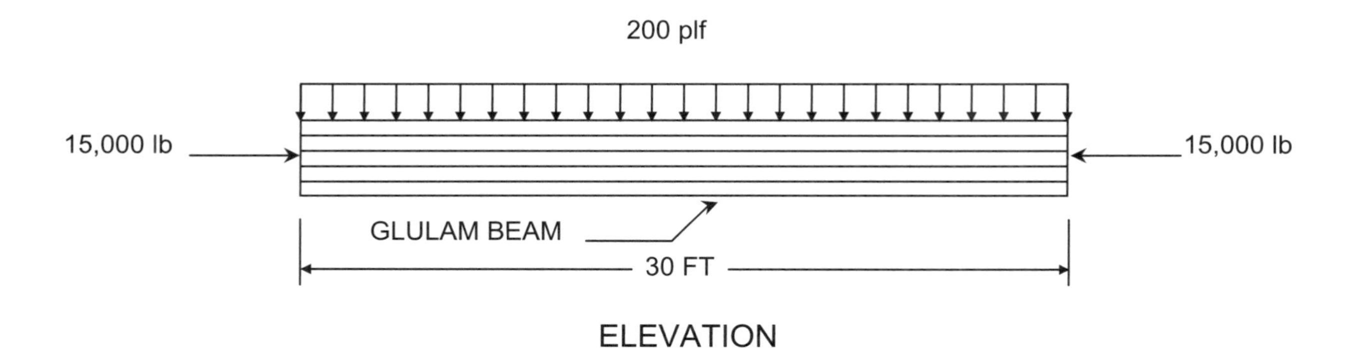

ELEVATION

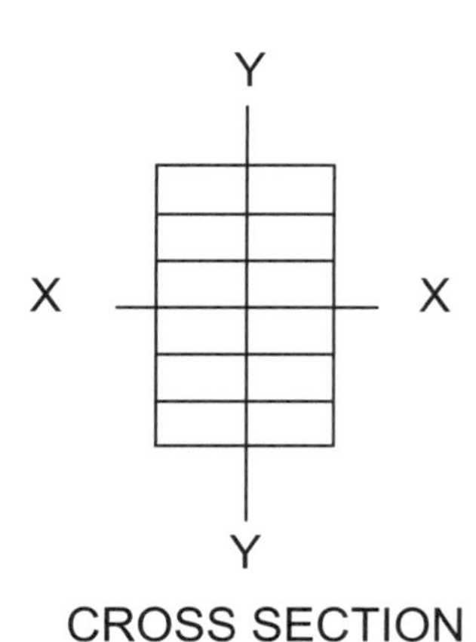

CROSS SECTION

Questions 514–518: The following summary of a geotechnical report applies to foundations and piles in **Figures 1 through 4**.

Foundations: Structures may be supported on spread footings with an allowable bearing pressure of 5,000 psf net (neglecting weights of overburden and concrete). A one-third increase in bearing pressure is allowed for wind and earthquake loads. Only 85% of dead loads may be used to resist wind or earthquake uplift loads. The minimum factor of safety is 1.5 for sliding and overturning.

Piles: Support is provided by combined end-bearing and surface friction. End-bearing capacity is 30,000 psf. Surface frictional resistance equals 1,500 psf in native soil. Neglect down-drag in fill soils at this site.

Lateral Earth Pressure: Design walls supporting earth for lateral earth pressure, using a triangular pressure distribution with the following equivalent fluid pressures:

	Above Water Table	Below Water Table
Active	27 pcf	21 pcf
At-rest	45 pcf	35 pcf
Passive	300 pcf	230 pcf

Assume the soil above the water table is saturated.

Design Data:

Concrete unit weight = 150 pcf (normal weight)
Soil unit weight = 120 pcf (saturated)

GO ON TO THE NEXT PAGE

STRUCTURAL AFTERNOON SAMPLE QUESTIONS

For **Questions 514 and 515**, refer to the square spread footing in **Figure 1**. The axial force P and moment M are transferred from the column to the footing and do **NOT** include the footing weight. The column shear force, if any, is neglected.

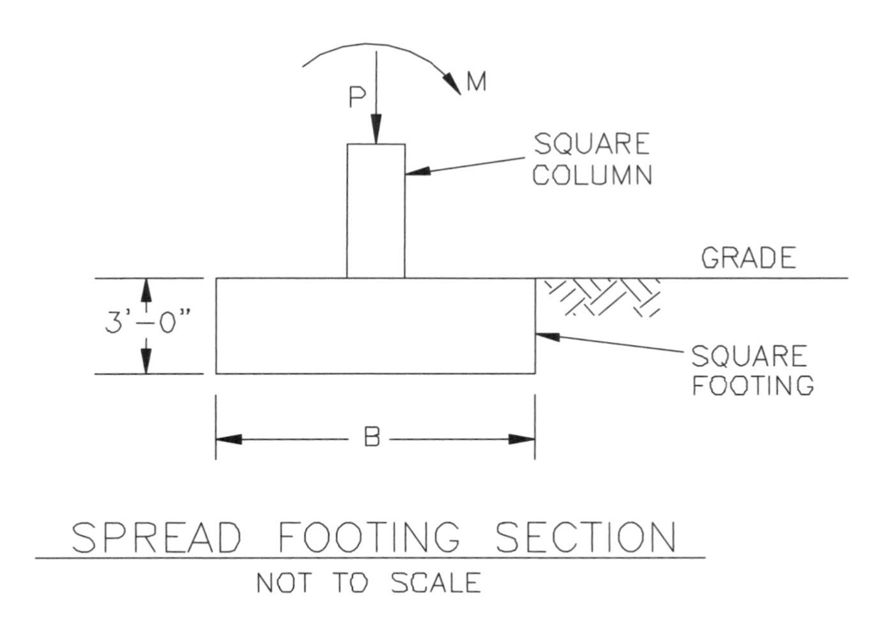

FIGURE 1

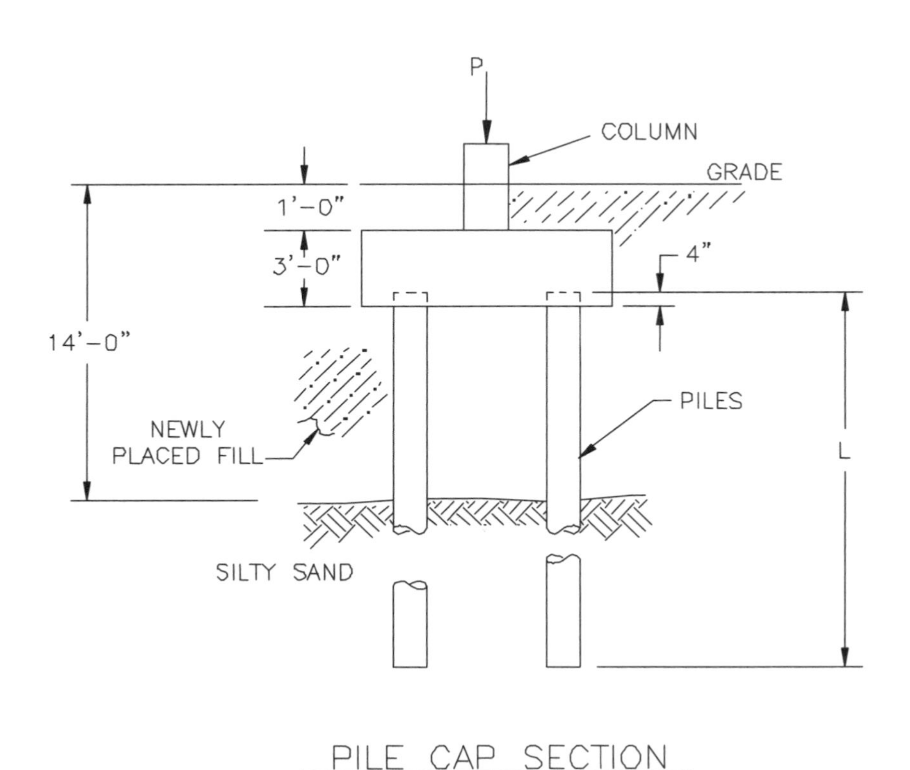

FIGURE 2

514. Assume the following service column axial forces and moments:

Service Load	Column Axial (kips) P	Column Moment (ft-kips) M
DL	400	50
Floor LL	150	20
Wind	±150	±150
Seismic	±250	0

The minimum dimension B (feet and inches) is most nearly:

(A) 11'-6"
(B) 12'-9"
(C) 13'-6"
(D) 14'-6"

515. Assume the footing dimension B is 8'-0" and the following service column axial forces and moment:

Service Load	Column Axial (kips) P	Column Moment (ft-kips) M
DL + LL	23	150

The actual maximum bearing stress (psf), including the footing weight on the soil, would be most nearly:

(A) unstable
(B) 4,000
(C) 2,600
(D) 2,000

516. For the pile cap shown in **Figure 2**, assume four symmetrically placed 14-inch-diameter piles and a total vertical force P of 1,000 kips, including the pile cap and backfill weights. The minimum length L (feet) for each pile is most nearly:

(A) 40
(B) 45
(C) 50
(D) 55

GO ON TO THE NEXT PAGE

For **Questions 517 and 518**, refer to the basement section shown in **Figure 3**. The water table is above the footings.

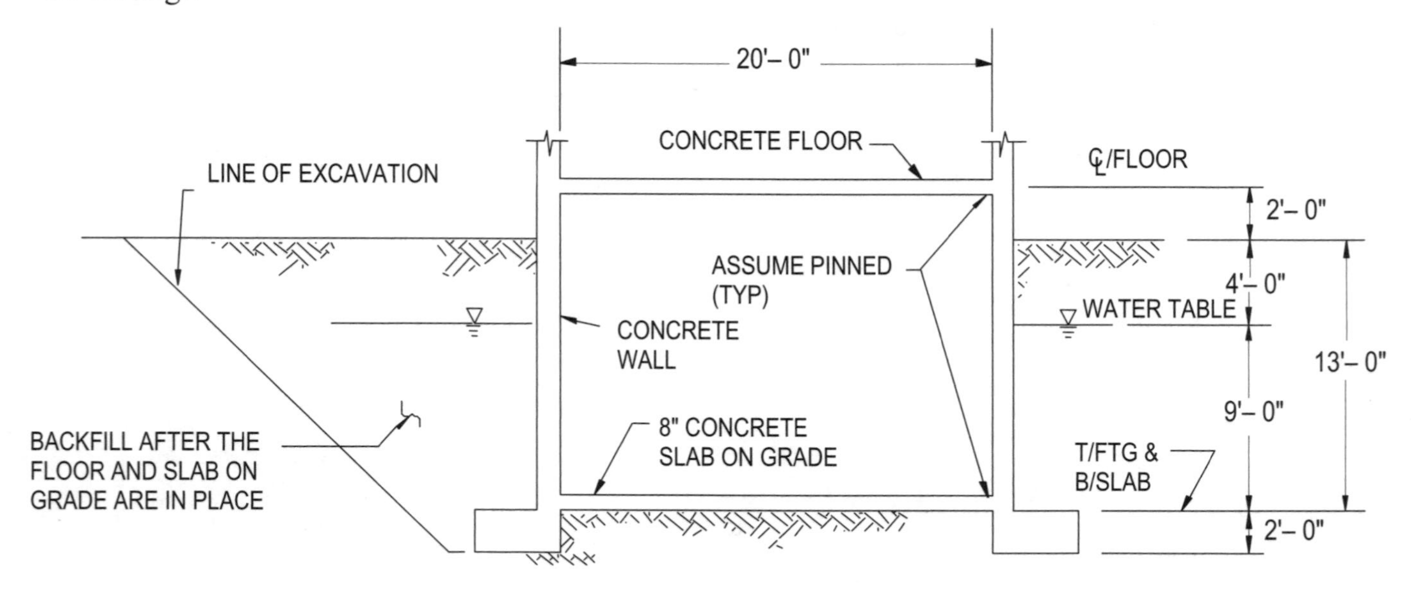

FIGURE 3

For **Question 518**, refer to the cantilever retaining wall system shown in **Figure 4**. The water table is below the footing.

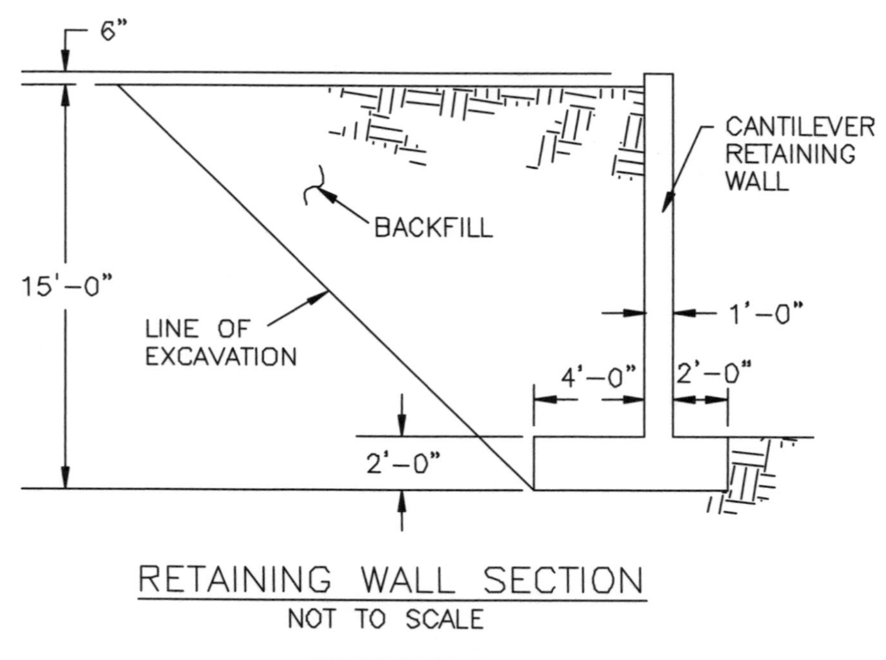

FIGURE 4

517. Considering only one of the basement walls and only that portion between the floor and the top of the footing, the total applied lateral force (lb/ft) is most nearly:

(A) 3,800
(B) 4,800
(C) 5,300
(D) 5,900

518. Assuming a surcharge equal to 2'-0" of soil, the total lateral force on the system (lb/ft) is most nearly:

(A) 3,050
(B) 3,850
(C) 5,050
(D) 6,400

GO ON TO THE NEXT PAGE

Structural Question 519 also appears as Geotechnical Question 520.

519. A segment of interstate highway requires the construction of an embankment of 500,000 yd^3. The embankment fill is to be compacted to a minimum of 90% of Modified Proctor maximum dry density.

A source of suitable borrow has been located for construction of the embankment. Assume that there is no soil loss in transporting the soil from the borrow pit to the embankment.

Use the following data:

Dry unit weight of soil in borrow pit	113.0 pcf
Moisture content in borrow pit	16.0%
Specific gravity of the soil particles	2.65
Modified Proctor optimum moisture content	13.0%
Modified Proctor maximum dry density	120.0 pcf

Assuming each truck holds 5.0 yd^3 and the void ratio of the soil is 1.30 during transport, the minimum number of truckloads of soil from the borrow pit that is required to construct the embankment is most nearly:

(A) 100,000
(B) 150,000
(C) 200,000
(D) 250,000

Structural Question 520 also appears as Transportation Question 508.

520. After purchasing a quarry and basic crushing equipment, the contractor is studying the following alternative, to improve the operation of the quarry. All alternative plans will produce equal amounts of crushed rock and equal revenue.

	Plan Number	
	Present	**Alternative**
First Cost ($)	0	10,000
Salvage ($)	0	1,000
Annual Cost ($)	250,000	248,000
Life (years)	—	5

The benefit-cost ratio of the alternative plan (using a 10% rate of return on investment) when compared to the present is most nearly:

(A) 0.6
(B) 0.8
(C) 1.0
(D) 1.2

TRANSPORTATION
AFTERNOON SAMPLE QUESTIONS

This book contains 20 transportation depth questions, half the number on the actual exam.

In some cases the same question appears in more than one of the depth modules because there is crossover of knowledge between the depth areas of civil engineering. This use of the same question on different depth modules could occur on the examination.

501. For the signalized intersection shown in the figure below, located in a flat suburban area with pedestrian demands that approach average flow rates, compute the minimum green time for North-South pedestrian intervals (G_p). The yellow clearance plus all red interval is 5.0 seconds. Assume average walking speeds. There are no separate pedestrian signals at the intersection.

The minimum green time for North-South pedestrian intervals, G_p (seconds), is most nearly:

(A) 22
(B) 27
(C) 29
(D) 32

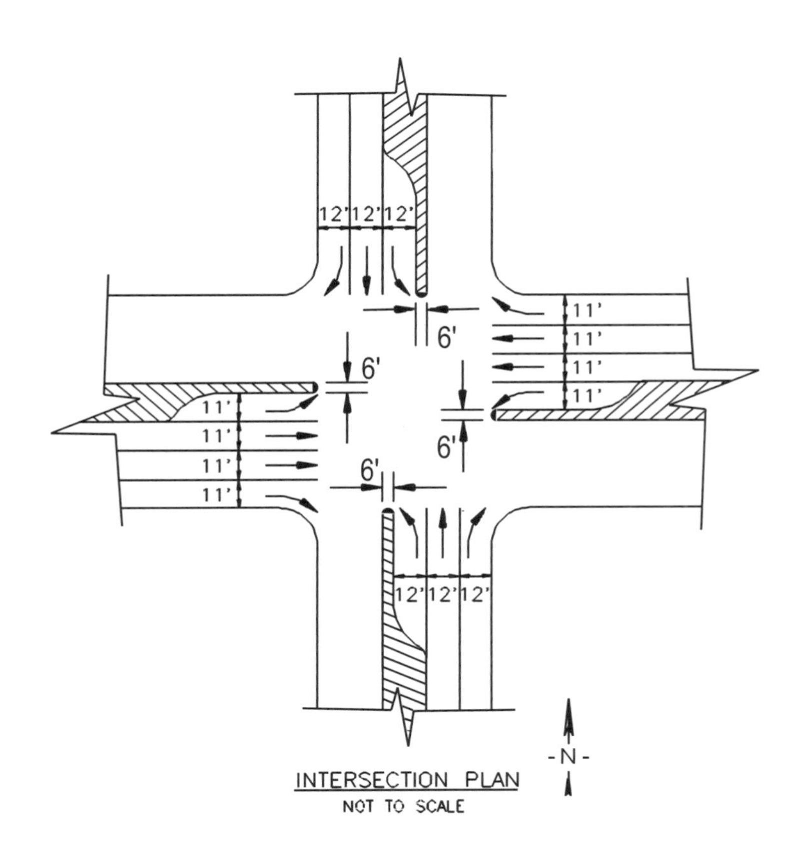

GO ON TO THE NEXT PAGE

502. Assuming the average spacing between the front bumpers of successive vehicles is approximately 30 feet, the best estimate for the "jam density" (vehicles/mile) in one lane of a typical urban arterial street is most nearly:

(A) 105
(B) 115
(C) 170
(D) 280

503. A traffic engineering study is being conducted on an urban four-lane arterial street. An automatic traffic counter using a pneumatic tube detector gave a total 1-year count of 24,560,000 axles at a continuous count station. Manual classification studies at the count station indicate that the traffic stream consisted of 85% passenger cars, 10% three-axle trucks, 3% four-axle trucks, and 2% five-axle trucks.

The average annual daily traffic (AADT) volume (vehicles) at this count station is most nearly:

(A) 2,800
(B) 25,770
(C) 27,400
(D) 30,310

504. The vertical alignment of a highway is shown in the figure below.

Roadway Data:
2-lane roadway
12-foot lanes

Assuming the vertical curve as shown with G_1 = –2.0% and G_2 = +6.0%, the stopping sight distance (feet) is most nearly:

(A) 440
(B) 490
(C) 525
(D) 650

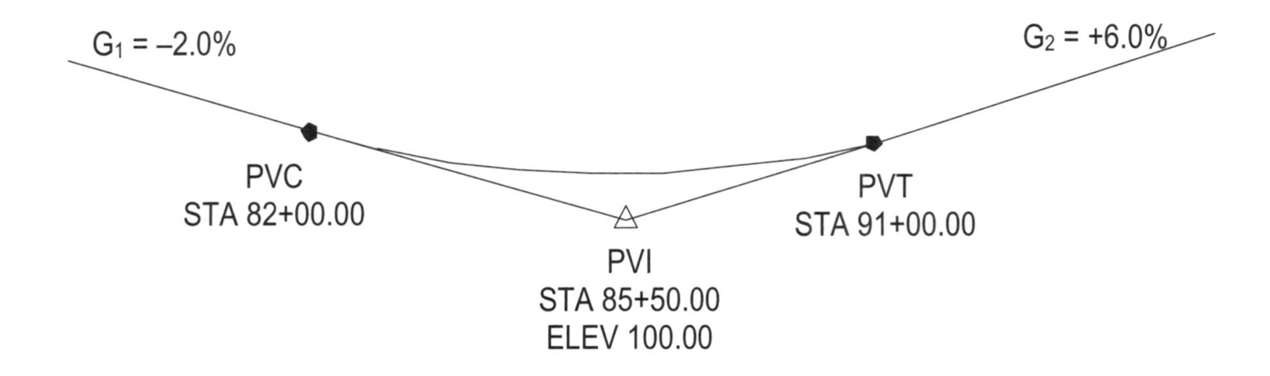

GO ON TO THE NEXT PAGE

505. Consider the relocation and construction of a section of rural highway. Data on costs are given in the table for the existing location and Alternative B. All values are in thousands of dollars. Use AASHTO guidelines and a 20-year analysis period with a 10% annual interest rate. Major maintenance will **NOT** be done in the 20th year.

The present value or present worth of the highway costs to the government, ignoring user costs, of Alternative B is most nearly:

(A) $6,000,000
(B) $6,150,000
(C) $6,350,000
(D) $6,600,000

	Existing	**Alternative B**
First Cost	None	$6,000
Annual Maintenance	$200	
Annual Maintenance		
1st 10 years		$50
2nd 10 years		$75
Major Maintenance		
Every 5 years		
Every 10 years		$300
Residual Value	$500	$3,000
Annual Road User Costs	$1,400	$660

TRANSPORTATION AFTERNOON SAMPLE QUESTIONS

506. If the traffic volume on a highway in 1994 is 30,000 vehicles per day, with a predicted annual growth rate of 5%, the traffic volume (vehicles per day) in the year 2000 will be most nearly:

(A) 46,000
(B) 44,000
(C) 42,000
(D) 40,000

507. A highway project requires a concrete mix. The design mix is proportional on the basis of (1:2.25:3.25) weight. Cement content was specified at 6.28 sacks per yd^3. The aggregates are SSD and have specific gravities of 2.65 for both the fine and the coarse. The specific gravity of the cement is 3.15.

The water/cement ratio (gallons per sack) of the concrete mix is most nearly:

(A) 3.6
(B) 4.2
(C) 4.7
(D) 5.2

GO ON TO THE NEXT PAGE

Transportation Question 508 also appears as Structural Question 520.

508. After purchasing a quarry and basic crushing equipment, the contractor is studying the following alternative, to improve the operation of the quarry. All alternative plans will produce equal amounts of crushed rock and equal revenue.

	Plan Number	
	Present	**Alternative**
First Cost ($)	0	10,000
Salvage ($)	0	1,000
Annual Cost ($)	250,000	248,000
Life (years)	—	5

The benefit-cost ratio of the alternative plan (using a 10% rate of return on investment) when compared to the present is most nearly:

(A) 0.6
(B) 0.8
(C) 1.0
(D) 1.2

509. A transition curve (spiral) with a degree of curvature (D) = 3°, and a length of spiral = 250 feet. The rate of change in degree of curve, D, (degrees per station) along the spiral is most nearly:

(A) 1.20
(B) 1.35
(C) 1.50
(D) 2.25

GO ON TO THE NEXT PAGE

The following information applies to **Questions 510 and 511**.

The tangent vertical alignment of a section of proposed highway is shown in the figure below.

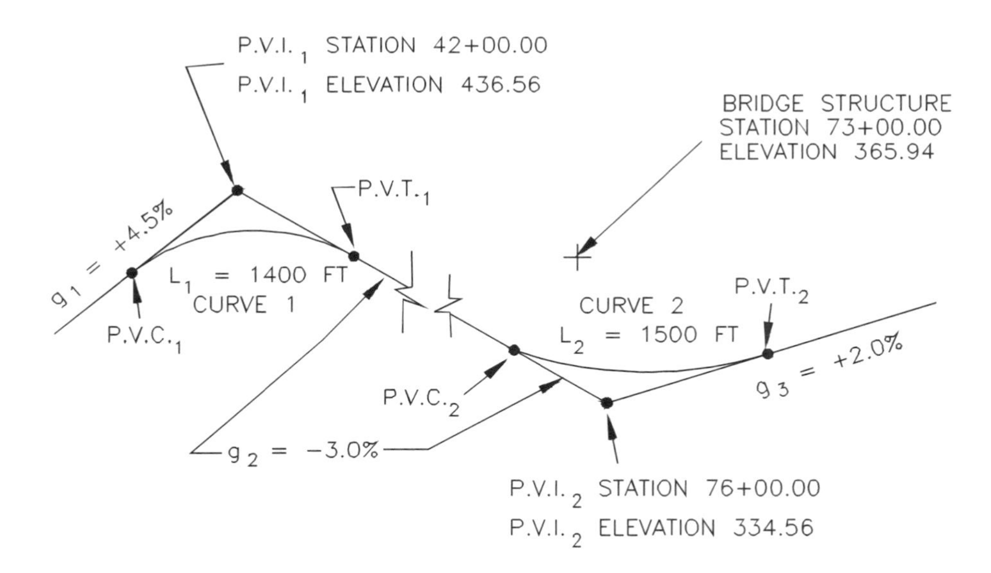

PROPOSED HIGHWAY VERTICAL ALIGNMENT

NOT TO SCALE

510. The station of the high point on Curve 1 is most nearly:

(A) 35 + 00
(B) 42 + 00
(C) 43 + 40
(D) 45 + 15

511. The vertical clearance (feet) between the bridge structure at Station 73 + 00 and the vertical curve is most nearly:

(A) 15.3
(B) 19.0
(C) 19.8
(D) 22.1

GO ON TO THE NEXT PAGE

512. A transition curve (spiral) is to be used to accomplish a change in cross-section from a normal crowned section to a fully superelevated section. The outer lane is to be gradually warped from the normal crowned section to a straight level section at the Tangent-to-Spiral (T.S.). The full superelevation is rotated about the centerline.

Degree of curve (D) = 3°
Design superelevation = 0.08 ft/ft
Grade = +1.00%
Crown = 0.015 ft/ft
Two 12-foot lanes
T.S. Station = 50 + 00.00
T.S. ℄ Elevation = 850.00 ft
Length of spiral = 250 ft

The station (feet) where full superelevation is reached is most nearly:

(A) 50 + 00
(B) 50 + 47
(C) 52 + 03
(D) 52 + 50

513. Assume a wet pavement coefficient of friction of 0.38, an approach speed of 45 mph, and a highway grade of –1.0% down to the tracks. Assume design values for perception-reaction time.

The stop line is located 15 feet from the nearside rail, and the driver is located 5 feet back from the front bumper of the vehicle. The required sight triangle distance (feet) for a vehicle to stop at the stop line for an approaching train is most nearly:

(A) 295
(B) 347
(C) 381
(D) 438

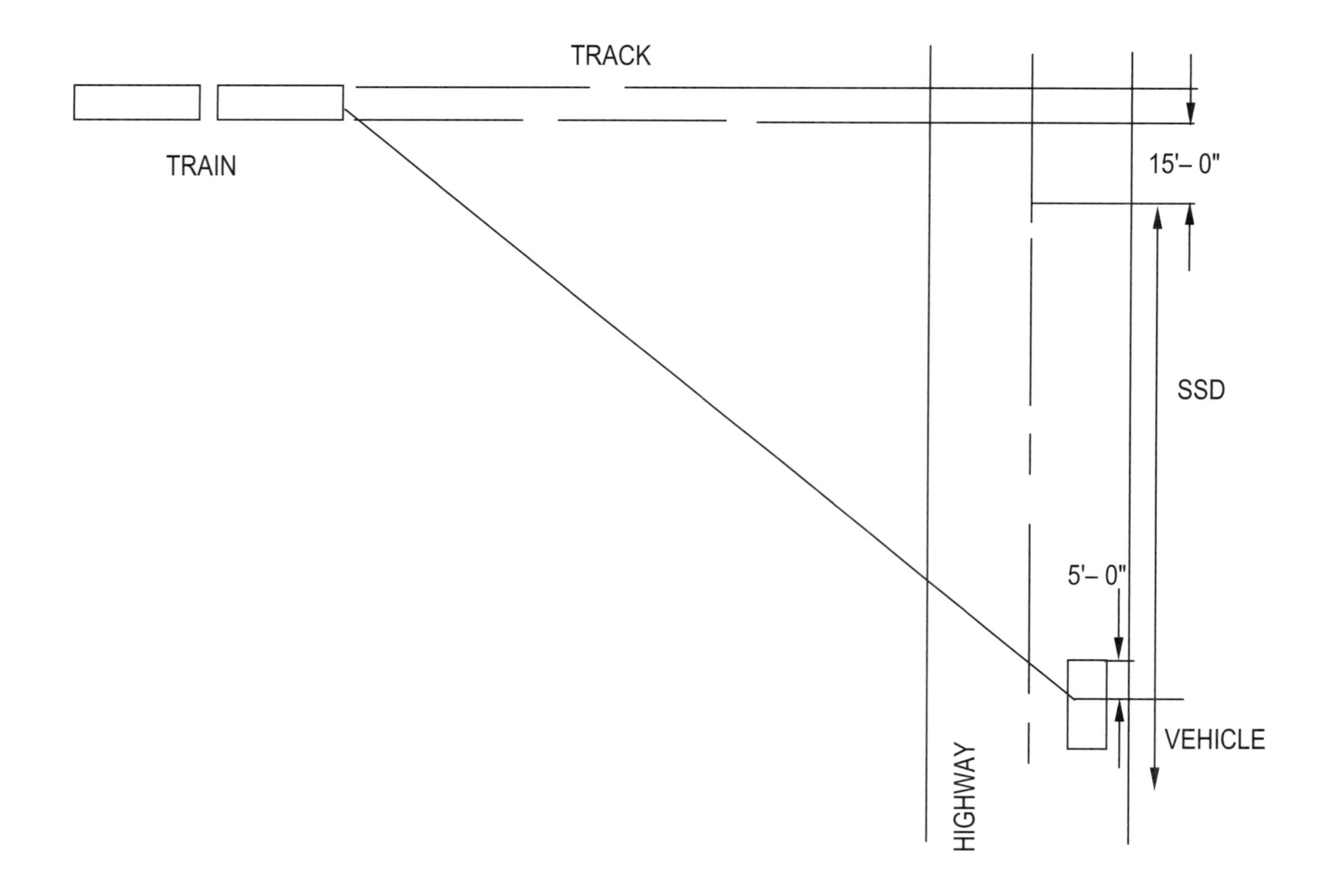

GO ON TO THE NEXT PAGE

514. Which statement is **NOT** true about the nuclear gauges used for measuring asphalt and soil compaction?

(A) To be an authorized user of a nuclear gauge an individual needs only to have a radiation-monitoring badge.

(B) Nuclear gauges can be used to read moisture as well as density.

(C) It is required that a nuclear gauge be transported in a properly labeled carrying case.

(D) Nuclear gauges use low-level radioactive material.

515. If the water-to-cement ratio is decreased for concrete, which statement is **NOT** true about concrete?

(A) Water tightness is decreased.
(B) Workability is decreased.
(C) Strength is increased.
(D) Durability is increased.

516. Which statement about the placement of asphalt is **NOT** true?

(A) The hotter the mix temperature above 300°F at the time of initial compaction, the better the results.

(B) In general, too low a percentage of air voids results in a shortened roadway life.

(C) Poor compaction will result if the mix is too cold at the time of initial compaction.

(D) The mix is usually in the temperature range of 260°F to 280°F at the time of spreading.

GO ON TO THE NEXT PAGE

517. The Manning roughness coefficient and average slope of a ditch adjacent to the road in the development area are 0.02 and 0.5%, respectively. The figure below depicts a cross section of the ditch. At a flow of 30 cfs, the depth of water in the ditch (inches) is most nearly:

(A) 6
(B) 9
(C) 12
(D) 15

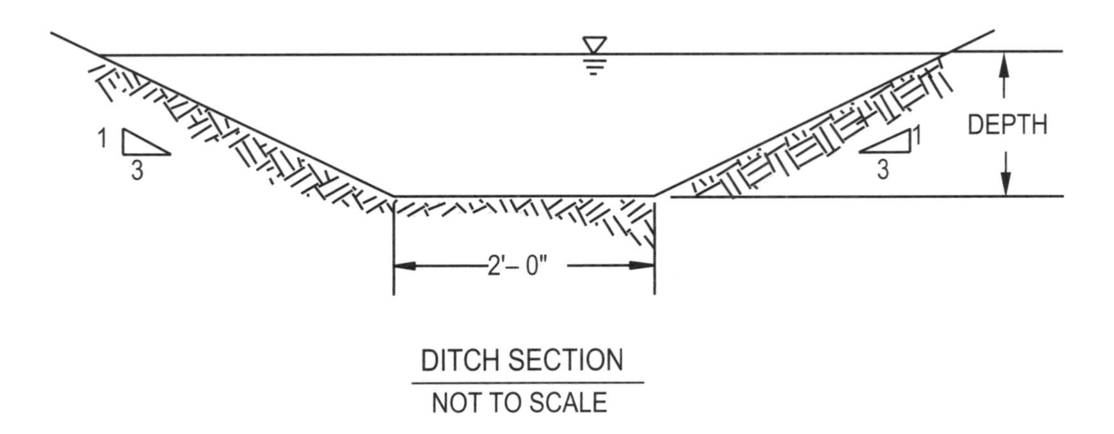

DITCH SECTION
NOT TO SCALE

518. A storm drain from a large shopping mall parking lot passes to a lake detention basin. The flow is 39 cfs with a flow depth to be equal to one-half the width. The channel width is 3.9 feet. The velocity of flow in the channel (fps) is most nearly:

(A) 3.2
(B) 4.1
(C) 5.3
(D) 6.2

519. A parcel of fair cover, undeveloped land (depicted in **Figure 1**) is to be developed into 1/4-acre-lot single-family housing. Historical rainfall data are contained in the table below. The land drains to an existing culvert as shown in **Figure 1**. Coefficient of runoff/rainfall intensity curves are provided in **Figure 2**.

Assuming a time of concentration of 7.5 minutes and using the data in **Figure 2**, the C value to be used for a 10-year storm for the proposed development would be most nearly:

(A) 0.75
(B) 0.79
(C) 0.82
(D) 0.85

Existing Condition—Fair Cover, Undeveloped Proposed Condition—1/4-Acre-Lot Single-Family Housing		
Time (minutes)	**Rainfall Data: (in/hr)**	
	10-year	**100-year**
5	2.95	5.10
10	2.08	3.48
20	1.44	2.38
60	0.72	1.30

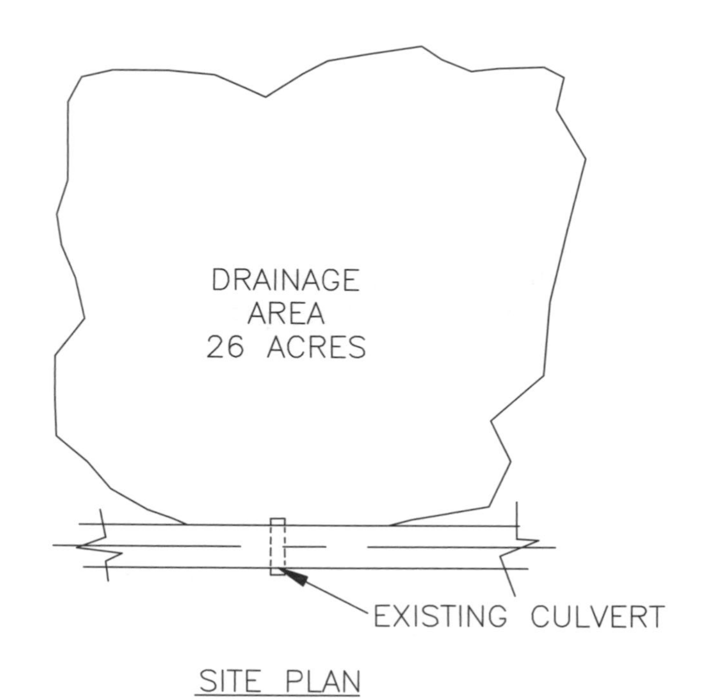

FIGURE 1

GO ON TO THE NEXT PAGE

The following figure applies to **Question 519**.

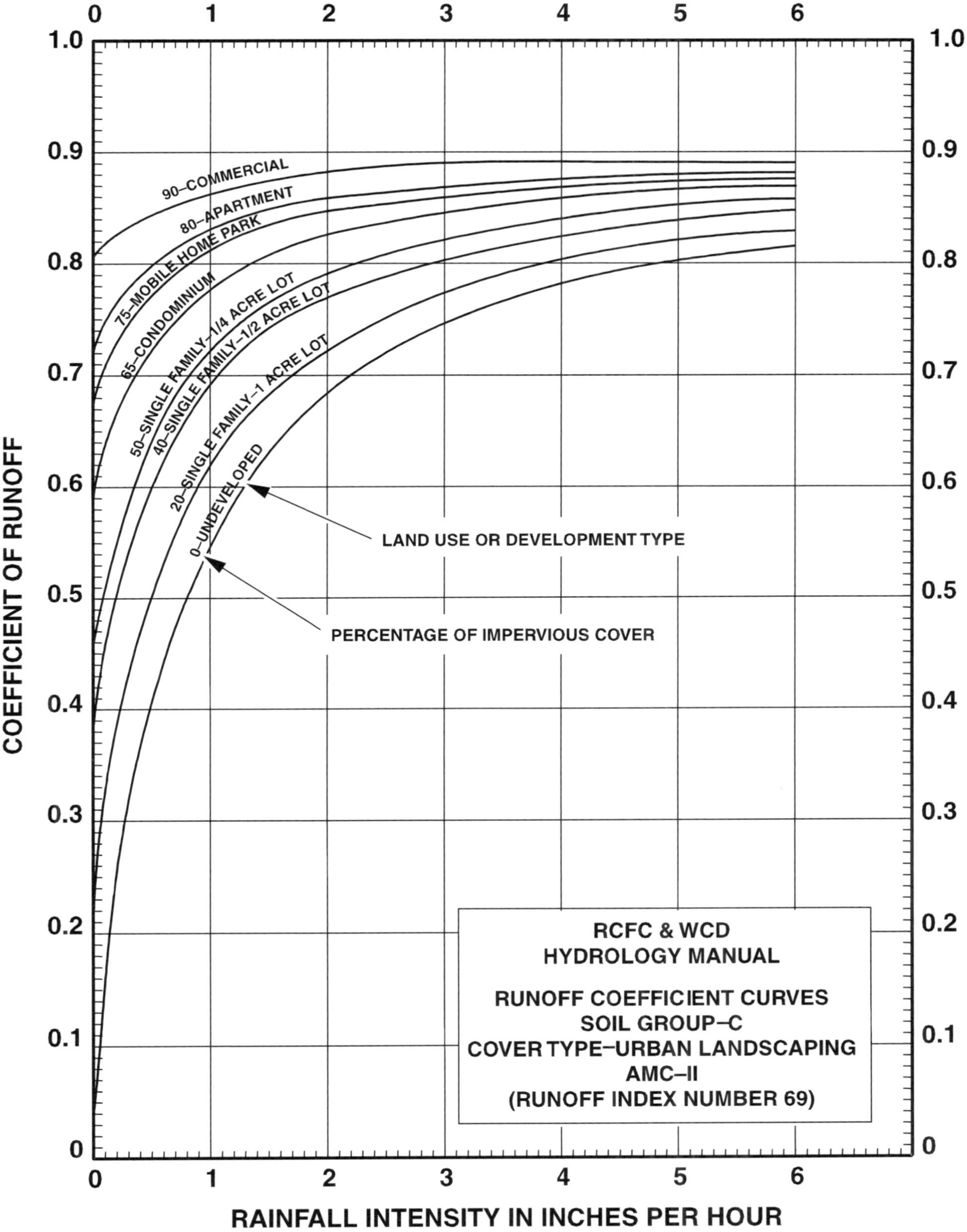
0
1
2
3
4
5
6
1.0
0.9
0.8
0.7
0.6
0.5
0.4
0.3
0.2
0.1
0
COEFFICIENT OF RUNOFF
90–COMMERCIAL
80–APARTMENT
75–MOBILE HOME PARK
65–CONDOMINIUM
50–SINGLE FAMILY–1/4 ACRE LOT
40–SINGLE FAMILY–1/2 ACRE LOT
20–SINGLE FAMILY–1 ACRE LOT
0–UNDEVELOPED
LAND USE OR DEVELOPMENT TYPE
PERCENTAGE OF IMPERVIOUS COVER
RCFC & WCD
HYDROLOGY MANUAL
RUNOFF COEFFICIENT CURVES
SOIL GROUP–C
COVER TYPE–URBAN LANDSCAPING
AMC–II
(RUNOFF INDEX NUMBER 69)
RAINFALL INTENSITY IN INCHES PER HOUR

FIGURE 2

GO ON TO THE NEXT PAGE

520. **Figure 1** depicts a drainage basin. **Figure 2** and the table provide hydrologic parameters for the basin.

Assume the velocity of water in all ditches is 1.7 fps. The peak runoff flow (cfs) from a 25-year storm at Point 2 in the basin, as predicted by the Rational Method, is most nearly:

(A) 6.2
(B) 7.5
(C) 9.8
(D) 13.3

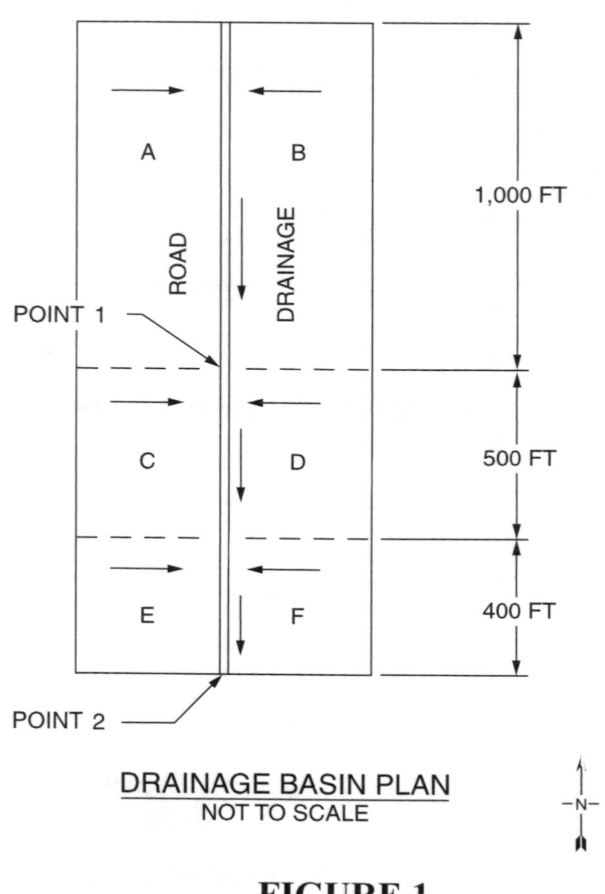

FIGURE 1

The following table and **Figure 2** apply to **Question 520** on the opposite page.

Identification	Land Use	Ave Residence/Acre	Area (Acres)	Runoff Coeff.	Time of Conc. (min)
A	Forest	0	8	0.1	30
B	Forest	0	12	0.1	35
C	Park/open space	0	6	0.15	25
D	Residential	2	6	0.4	20
E	Residential	2	4	0.4	15
F	Residential	4	4	0.6	10

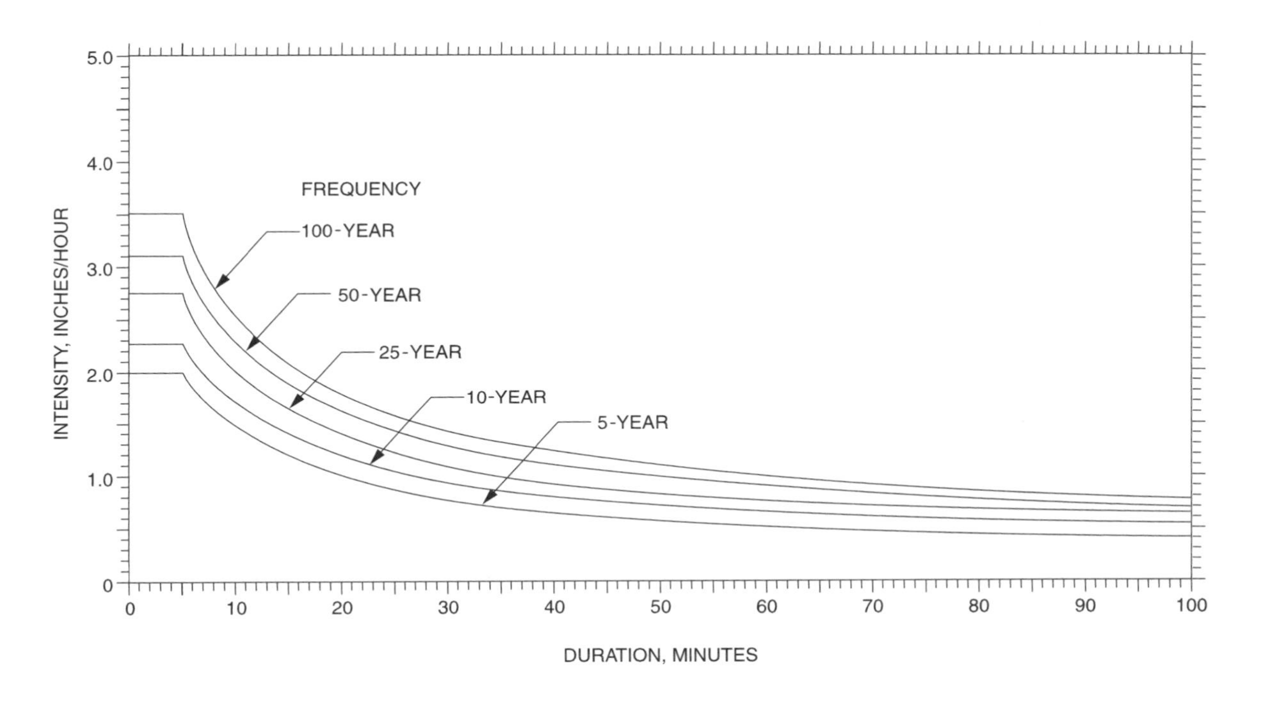

RAINFALL INTENSITY, DURATION, FREQUENCY

FIGURE 2

WATER RESOURCES
AFTERNOON SAMPLE QUESTIONS

This book contains 20 water resources depth questions, half the number on the actual exam.

In some cases the same question appears in more than one of the depth modules because there is crossover of knowledge between the depth areas of civil engineering. This use of the same question on different depth modules could occur on the examination.

501. Water flows into an old cast-iron, 12-inch-diameter distribution header at 1,000 gpm. A cross-over piping arrangement connects this header to an 8-inch auxiliary header, with exit flows as shown in the figure.

The pressure at Point A is given as 50-psi gage, the elevation at Point A is 110 feet, and the elevation of Point D is 90 feet. The pressure head at Point D is to be determined. Which of the following statements is most correct? Assume that the velocity heads can be neglected.

(A) The pressure at Point D is independent of the flow rate in pipe AD.

(B) The pressure at Point D is always less than the pressure at Point A.

(C) If the final flow rates in each pipe were given when the flows in the loop were completely balanced, the pressures at Point B and C could also be obtained with the given information.

(D) The pressure head at Point D is equal to the total energy at Point A minus the friction loss between Points A and D and minus the elevation energy at Point D.

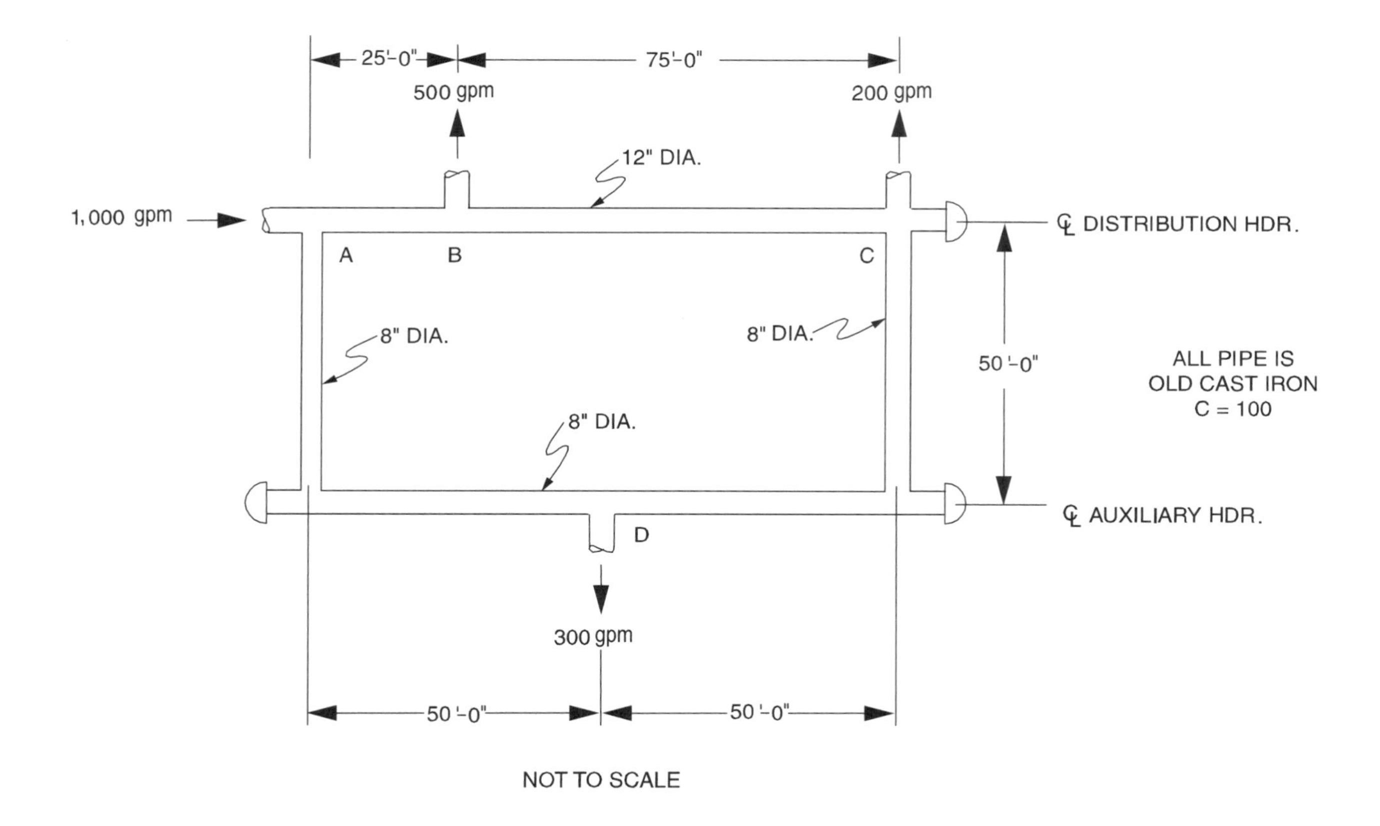

502. A valve is at the end of a 1,000-foot-long, 8-inch-diameter ductile iron water main. The pipe wall thickness is 0.27 inch, the bulk modulus for water is 3×10^5 psi, and the modulus of elasticity of the pipe material is 24×10^6 psi.

The minimum time (seconds) in which the valve can be closed without causing water hammer is most nearly:

(A) 0.5
(B) 1.0
(C) 1.5
(D) 2.0

GO ON TO THE NEXT PAGE

For **Questions 503–504**, the peak and minimum sanitary sewage flows in an interceptor are 6 MGD and 0.70 MGD, respectively. The sewer is constructed of vitrified clay in very good condition ($n = 0.011$). Assume that n is constant with depth. Assume that the pipe slope is 0.001.

503. The most economical pipe size (inches) to prevent surcharged conditions is most nearly:

(A) 12
(B) 18
(C) 24
(D) 30

504. For the minimum flow of 0.70 MGD, the depth (inches) of flow in a 36-inch sewer ($n = 0.011$, constant with depth) is most nearly:

(A) 5
(B) 7
(C) 8
(D) 9

For **Questions 505–506**, a pump in **Figure 1** with the characteristics given in **Figure 2** is to deliver water through the existing system indicated in the figure. Assume all elevations are constant. The Hazen-Williams equation is to be used to estimate friction losses. The total length of pipe is 3,000 feet and made of cast iron with a diameter of 10 inches and a Hazen-Williams coefficient of 100. Elevations are indicated on **Figure 1**.

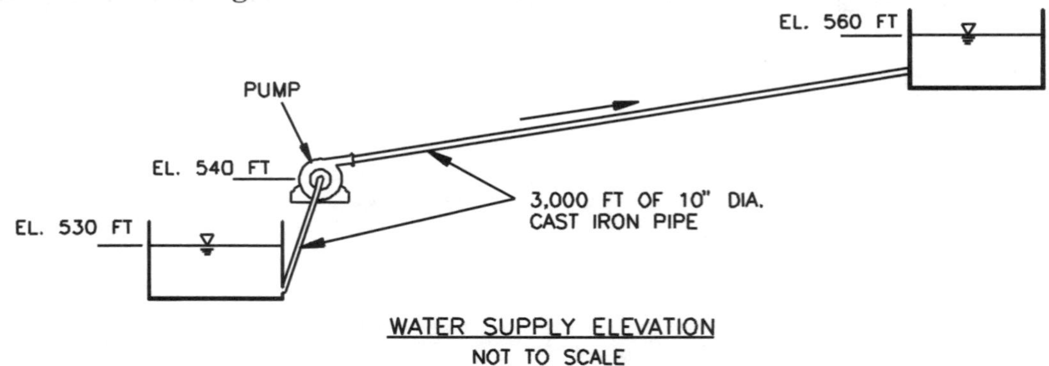

FIGURE 1

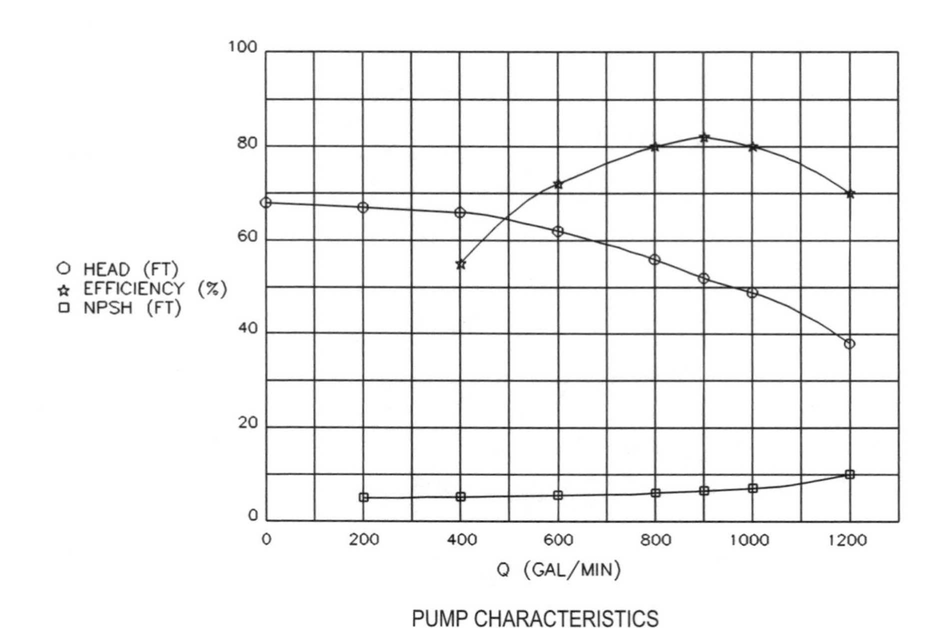

PUMP CHARACTERISTICS

FIGURE 2

GO ON TO THE NEXT PAGE

505. If two of these pumps were installed in parallel in the system, the flow rate (gpm) would be most nearly:

(A) 900
(B) 1,050
(C) 1,200
(D) 1,550

506. Assuming the suction line length is 30 feet, the maximum elevation (feet) of the pump to avoid cavitation when operating at its point of maximum efficiency will be most nearly:

(A) 555
(B) 547
(C) 534
(D) 524

GO ON TO THE NEXT PAGE

507. A pump is to deliver water through the existing system as indicated in the figure below.

Which of the following statements regarding valves in the system are true?

I. Flow control valves should not be located on the suction side of the pump.

II. A globe valve installed as a flow regulation valve would permit more flow through the system than a gate valve.

III. A check valve should be installed on the suction side of the pump.

(A) I only
(B) II only
(C) I and II only
(D) I and III only

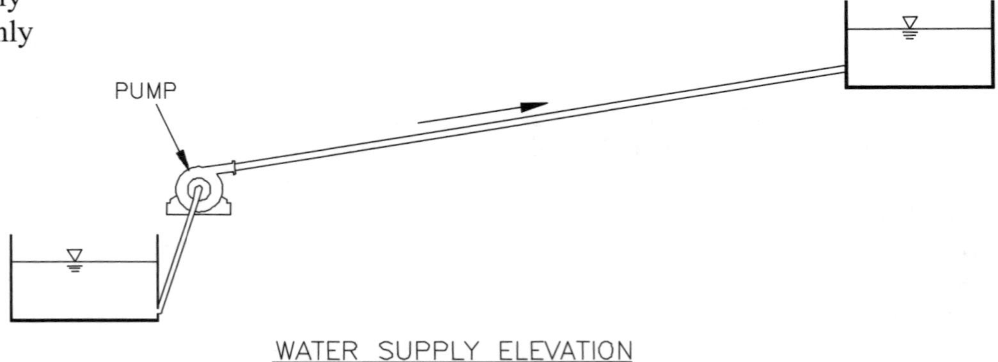

WATER SUPPLY ELEVATION
NOT TO SCALE

508. Assume the SCS curve number for an area is 50 and that precipitation during a 24-hour storm was 5 inches. The average depth of runoff water (inches) from the area is most nearly:

(A) 0.7
(B) 1.2
(C) 2.5
(D) 3.1

For **Questions 509–511**, **Figure 1** depicts a drainage basin, and **Figure 2** and the table below provide hydrologic parameters for the basin.

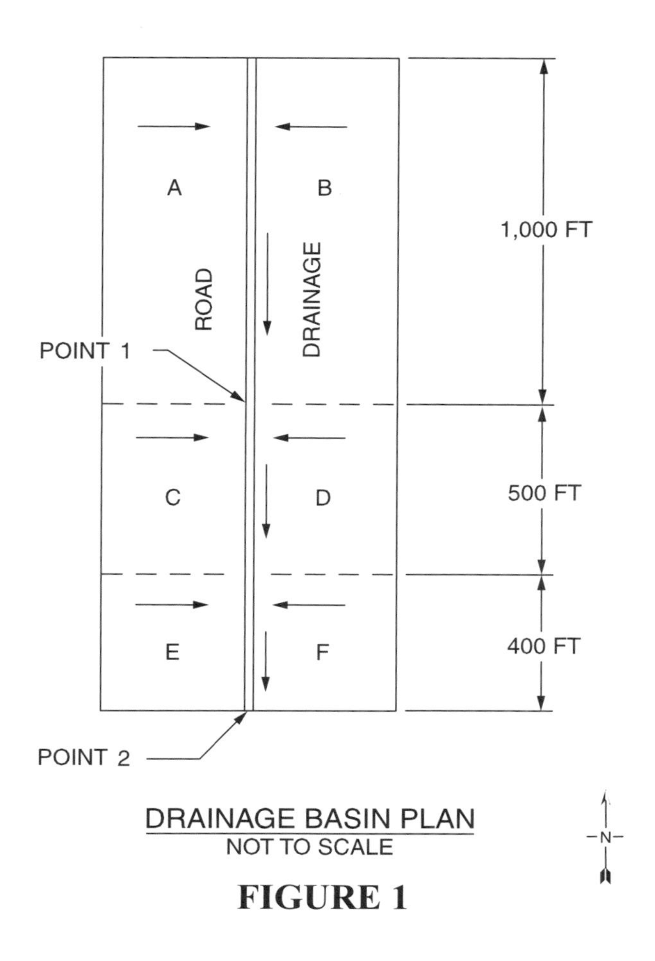

FIGURE 1

Identification	Land Use	Ave Residence/Acre	Area (Acres)	Runoff Coeff.	Time of Conc. (min)
A	Forest	0	8	0.1	30
B	Forest	0	12	0.1	35
C	Park/open space	0	6	0.15	25
D	Residential	2	6	0.4	20
E	Residential	2	4	0.4	15
F	Residential	4	4	0.6	10

GO ON TO THE NEXT PAGE

The following figure applies to **Question 509**.

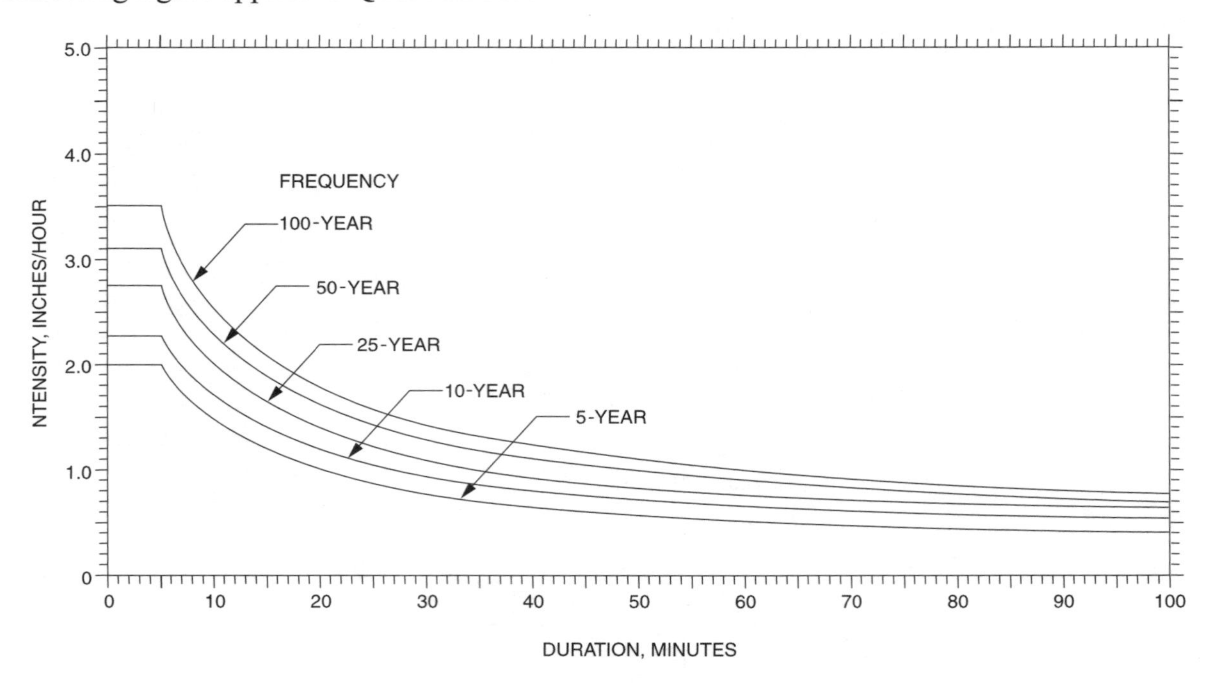

RAINFALL INTENSITY, DURATION, FREQUENCY

FIGURE 2

509. Assume the velocity of water in all ditches is 1.7 fps. The peak runoff flow (cfs) from a 25-year storm at Point 1 in the basin, as predicted by the Rational Method, is most nearly:

(A) 1.7
(B) 2.0
(C) 2.3
(D) 3.9

GO ON TO THE NEXT PAGE

510. The following discharges were measured at Point 2 of the drainage basin.

Hour	Stormwater Discharge (cfs)
0	0
2	2.5
4	5
6	7.5
8	4.0
10	2.0
12	1.0
14	0

For a storm of the same duration, the unit hydrograph discharge (cfs) at Hour 10 would be most nearly:

(A) 1.1
(B) 1.8
(C) 2.5
(D) 3.0

511. The runoff coefficient representative of the entire drainage basin is most nearly:

(A) 0.10
(B) 0.15
(C) 0.25
(D) 0.30

The following information and **Figures 1 and 2** apply to **Question 512.**

A pump in **Figure 1** with the characteristics given in **Figure 2** is to deliver water through the existing system indicated in the figure. The Hazen-Williams equation is to be used to estimate friction losses. The total length of pipe is 3,000 feet and made of cast iron with a diameter of 10 inches and a Hazen-Williams coefficient of 100. Elevations are indicated on **Figure 1**.

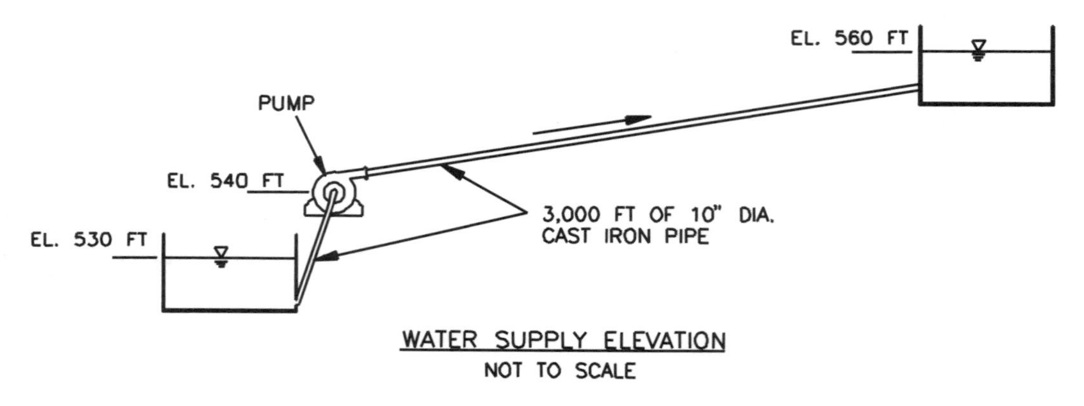

FIGURE 1

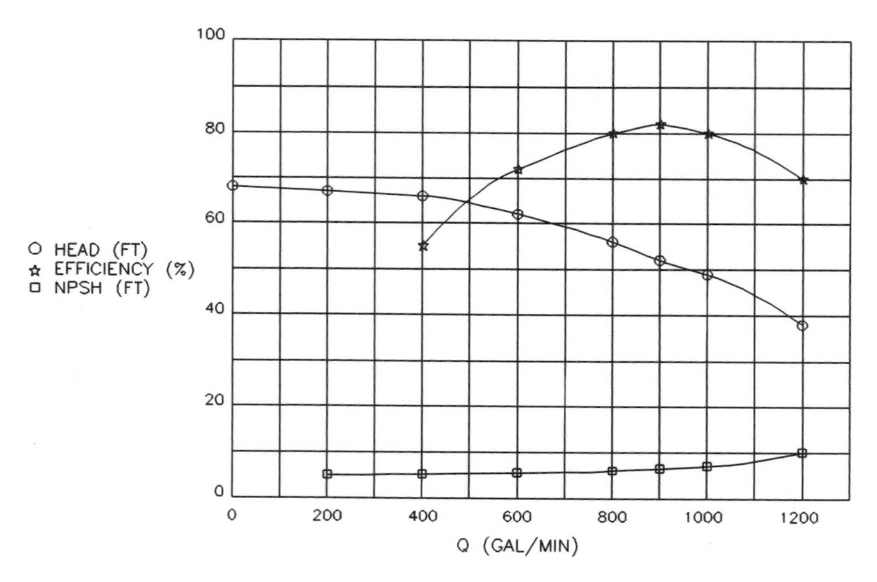

FIGURE 2

512. The flow (gpm) through the existing system will be most nearly:

(A) 500
(B) 800
(C) 900
(D) 1,030

513. A 6.0-MGD water treatment plant is being planned that will use a local river as its water source. The rapid-mix units will be designed for a 30-second detention time. The maximum size of each unit will be 50 ft^3. The number of units and volume (ft^3), respectively, needed of rapid mixers is most nearly:

(A) 6, 50
(B) 6, 46.4
(C) 5, 60
(D) 5, 65.4

514. A wastewater treatment plant designed for a community of 30,000 people has a flow of 100 gal/person/day and a BOD_5 of 0.2 lb/person/day. The upstream characteristics of the receiving waters are as follows:

Flow = 15 cfs

BOD_5 = 1.5 mg/L

The water quality standards require the in-stream BOD_5 to be less than 7 mg/L in this area. The minimum BOD removal efficiency that the treatment plant should achieve is most nearly:

(A) 90%
(B) 85%
(C) 80%
(D) 75%

GO ON TO THE NEXT PAGE

515. Disinfection is to be added to a sewage treatment plant with an average daily flow of 13 MGD, and an hourly peaking factor of 2.5. The average coliform count (N_o), 10,000 org/100 ml, must be reduced to 200 org/100 ml (N_t).

The planned facilities shown in the figure below include a gas chlorinator, a horizontal channel for a hydraulic jump for mixing, and a detention basin.

Codes require a 15-minute contact time for peak hourly flow, a 30-minute contact time for average daily flow, and a maximum chlorine dosage capability of 15 mg/L.

The equation

$$\frac{N_t}{N_o} = (1 + 0.23\, c_t t)^{-3}$$

describes the chlorination process. Units for c_t and t are mg/L and minutes, respectively.

The required total chlorine residual (mg/L) after 15 minutes of contact time is most nearly:

(A) 0.78
(B) 1.11
(C) 1.43
(D) 1.87

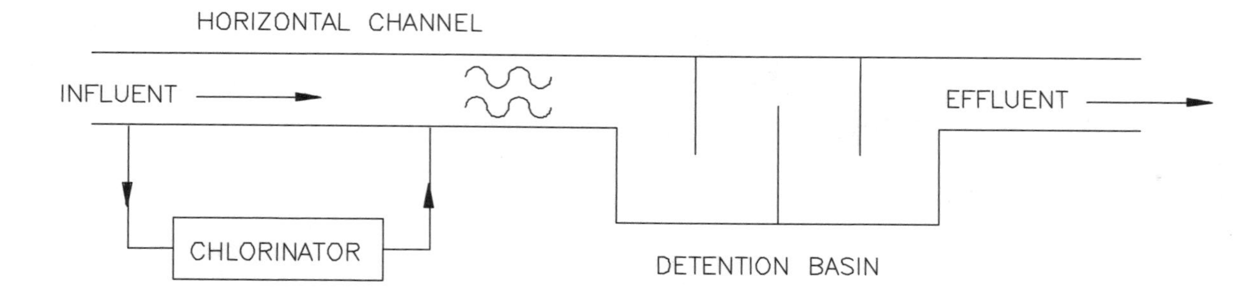

516. At an industrial complex using ammonia for refrigeration, a leak was discovered from the supply pipe of ammonia to a nearby stream.

Water samples were collected 10 miles downstream from the point of discharge and analyzed. At the downstream location, the peak ammonia concentration was 10-mg/L ammonia-nitrogen as N, the pH averaged 7.5, and the water temperature averaged 77°F.

Assume the pKa for NH_3 is 9.3. The concentration (mg/L) of nonionized ammonia nitrogen (NH_3 - N) 10 miles downstream from the point of release is most nearly:

(A) 0.05
(B) 0.10
(C) 0.15
(D) 0.20

517. A wastewater treatment plant discharges a secondary-treated effluent into a receiving stream.

Assume the following characteristics of the wastewater/river mixture at the point of discharge:

Ultimate BOD	=	10 mg/L
Dissolved oxygen	=	6 mg/L
Temperature	=	20°C
DO_{sat}	=	9.08 mg/L
River velocity	=	1 fps
Reaeration rate constant (base e)	=	0.40 day^{-1}
Deoxygenation rate constant (base e)	=	0.23 day^{-1}

Also, assume that no other wastewater sources are discharged into the river.

The location (miles) of the critical dissolved oxygen concentration from the point of wastewater discharge is most nearly:

(A) 0.0
(B) 11.0
(C) 28.5
(D) 76.0

518. A 16-inch-diameter drinking water well is located in a local confined aquifer. The aquifer formation is composed of well-sorted, loosely cemented, fine-to-coarse-grained sand. In addition, the top of the aquifer is located 1,450 feet below the ground surface and is 85 feet thick. In a recent pumping test, it was determined that the transmissivity was 38,000 gpd per foot and the storativity was 3×10^{-4}. The well fully penetrates the aquifer.

Assuming the piezometric level was 1,400 feet below the ground surface prior to pumping and is 1,420 feet below the ground surface at the well casing after 100 days of pumping, the pumping rate of the well (gpm) is most nearly:

(A) 290
(B) 310
(C) 350
(D) 1,010

Water Resources Question 519 also appears as Environmental Question 514.

519. A soil sample was tested in the apparatus as shown in the following figure.

The coefficient of permeability (inches per minute) of the soil sample is most nearly:

(A) 0.02
(B) 0.14
(C) 0.21
(D) 0.44

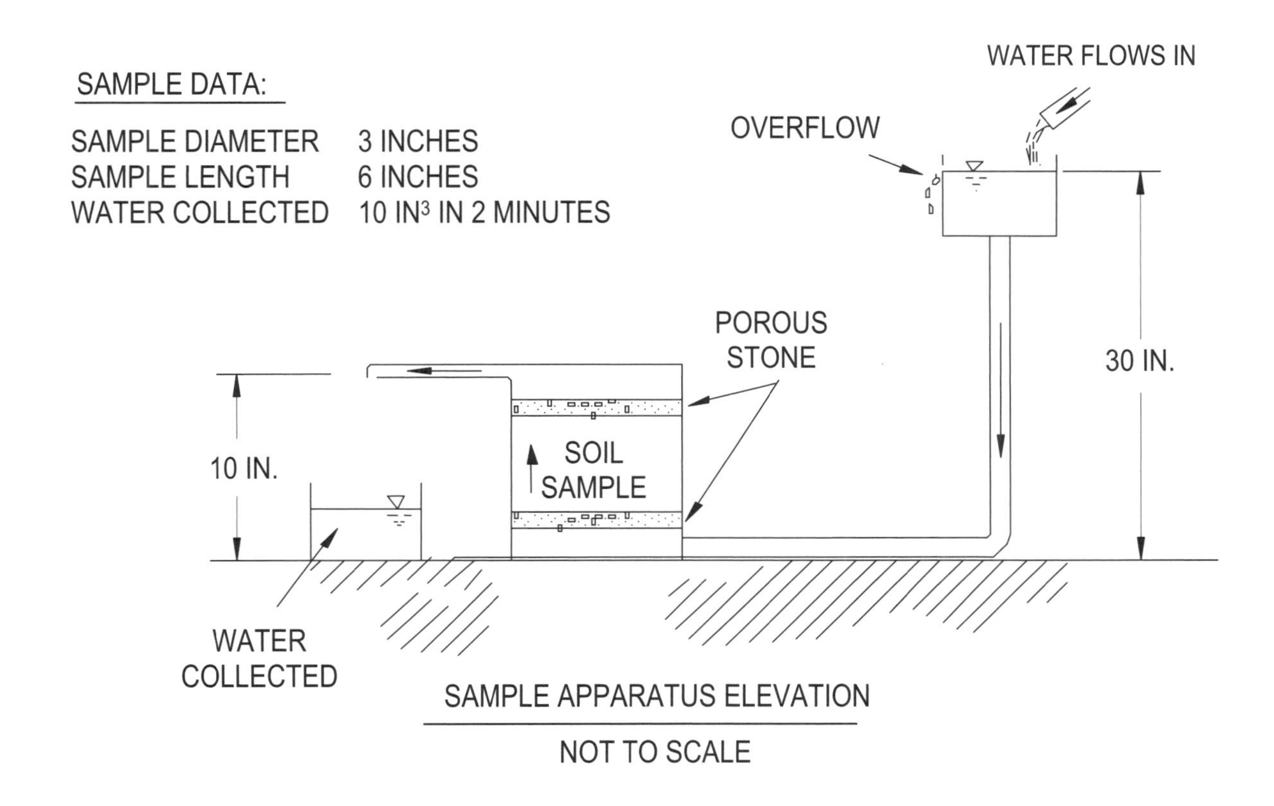

GO ON TO THE NEXT PAGE

Water Resources Question 520 also appears as Environmental Question 515.

520. A soil sample was tested in the apparatus as shown in the following figure.

The coefficient of permeability (inches per minute) of the soil sample is most nearly:

(A) 0.010
(B) 0.013
(C) 0.018
(D) 0.023

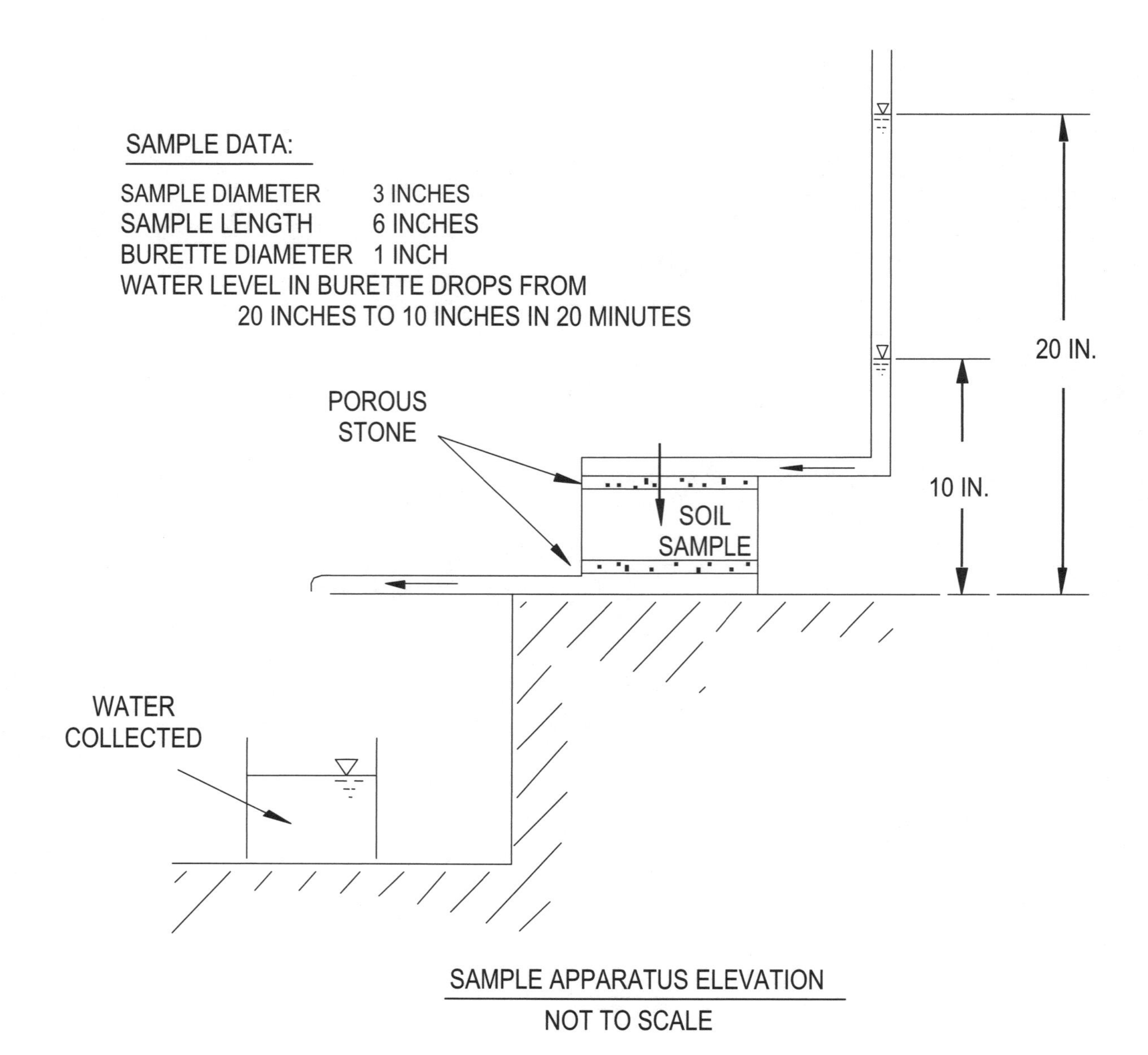

SAMPLE APPARATUS ELEVATION

NOT TO SCALE

CIVIL BREADTH
MORNING SAMPLE SOLUTIONS

CORRECT ANSWERS TO THE CIVIL MORNING SAMPLE QUESTIONS

Detailed solutions for each question begin on the next page.

101	B
102	C
103	B
104	B
105	B
106	B
107	C
108	B
109	B
110	D
111	B
112	D
113	C
114	C
115	B
116	B
117	B
118	A
119	A
120	D

CIVIL MORNING SAMPLE SOLUTIONS

101. Removal efficiency = 80%

$\therefore$ % Remaining = 20%

$\therefore$ kt from the Fig = 2.5

Calculate the rate constant under winter conditions:

$$k_{15} = k_{20}\left(\theta^{(15-20)}\right) = 0.187 \text{ day}^{-1}$$

Now determine the detention time $t = \frac{kt}{k} = \frac{2.5}{0.187} = 13.4$ days

$\therefore$ Minimum volume = Q × t = 13.4 mg

THE CORRECT ANSWER IS: (B)

102. THE CORRECT ANSWER IS: (C)

103. Determine the energy content in Btu/lb of MSW before recycling.

Component	% by Wt. lb/100 lb	Btu/lb	% Recycle	Btu before Recycling	Btu after Recycling
Food Waste	8.0	2,500	0.0	20,000	20,000
Paper	39.0	6,700	50.0	261,300	130,650
Cardboard	6.0	7,000	75.0	42,000	10,500
Plastics	5.0	15,000	25.0	75,000	56,250
Textiles	2.5	7,200	0.0	18,000	18,000
Rubber	1.0	10,500	0.0	10,500	10,500
Leather	0.3	7,500	0.0	2,250	2,250
Yard Wastes	20.5	3,000	25.0	61,500	46,125
Wood	3.0	8,000	25.0	24,000	18,000
Glass	6.0	75	0.0	450	450
Metals	5.7	200	0.0	1,140	1,140
Dirt, ash, etc.	3.0	3,000	0.0	9,000	9,000
	100.0			525,140	322,865

$\therefore$ Reduction in energy content $= \frac{525{,}140 - 322{,}865}{525{,}140} = 38.5\% = 40\%$

THE CORRECT ANSWER IS: (B)

104. Determine minimum bottom elevation of containment cell.

The highest groundwater elevation within cell area = groundwater elevation at Point A less the groundwater drop to closest point of the cell, Point C highest groundwater elevation within cell area = groundwater elevation at Point A less the (distance A to C)*i

Distance Well A to cell = $[(500)*(500) + (100)*(100)]^{0.5}$ = 510 feet

Highest groundwater elevation within cell area = 229.75 – 510*0.0049 = 227.25

Minimum bottom elevation of containment cell = highest groundwater elevation within cell area plus 5 feet

Minimum bottom elevation of containment cell = 227.25 + 5 feet

Minimum bottom elevation of containment cell = 232 feet

THE CORRECT ANSWER IS: (B)

105. Determine the effective overburden pressure.

The depth below the original ground surface to the center of the clay layer is 15 ft (5 ft of moist sand, 5 ft of saturated sand, and 5 ft of clay) which is 10 ft below the water table.

σ'_o = weight of soil above a depth of 15 ft minus the weight of water below the water table

$= \Sigma\, \gamma \Delta h - \gamma_w\, h_w$

$\sigma'_o = (115)\,(5.0) + (130)\,(5.0) + (95)\,(5.0) - (62.4)\,(10.0) = 1{,}700 - 624$

$\sigma'_o = 1{,}076$ psf

THE CORRECT ANSWER IS: (B)

106. Determine the void ratio of the sand.

$$e = \frac{G\gamma_w}{\gamma_D} - 1 = \frac{(2.65)(62.4)}{107} - 1$$

e = 0.55 which is most nearly 0.5

THE CORRECT ANSWER IS: (B)

107. Determine the volume of water to be added to the soil from the borrow pit.

NOTE: The solution must be based on dry, not moist, unit weights.

The weight of water to be added is

$W_w = (500{,}000 \text{ yd}^3)(27 \text{ ft}^3/\text{yd}^3)(0.90)(116.0 \text{ pcf})(0.05) = 70.47 \times 10^6 \text{ lb}$

The volume of water added is

$V_w = (70.47 \times 10^6 \text{ lb})/(8.33 \text{ lb/gal})$

$= 8.46 \times 10^6$ gal, say 8,500,000

THE CORRECT ANSWER IS: (C)

108. Determine the active lateral earth pressure coefficient.

$$K_A = \tan^2\left(45 - \frac{\phi}{2}\right) = \tan^2\left(45 - \frac{28}{2}\right)$$

$K_A = 0.36$ which is most nearly 0.35

THE CORRECT ANSWER IS: (B)

109. $P = 6(54) + 12(15 + 40) = 984$ plf

THE CORRECT ANSWER IS: (B)

110. Assuming the approximate wall height and earth height to be the same,

$L = 8.50$ ft
$p =$ hydrostatic pressure at the bottom of the wall
$= (40)(8.50) = 340$ psf

Since the wall is supported on both ends, it will act as a simply-supported beam with a span of 8.5 ft. As the figure shows, for triangularly distributed load the maximum moment will occur $0.5774 \times 8.5 = 4.91$ feet from the top of the wall.

$$M_{max} = [(481.7)(4.91)] - (4.91)^2(196.4/6)$$

$$= 2{,}365.1 - 789.1 = 1{,}576 \text{ ft-lb/ft}$$

or

$$M_{max} = 0.1283\ WL$$

$$W = 340(8.5)/2 = 1{,}445 \text{ lb/ft}$$

$$M_{max} = 0.1283(1{,}445)(8.5) = 1{,}576 \text{ ft-lb/ft}$$

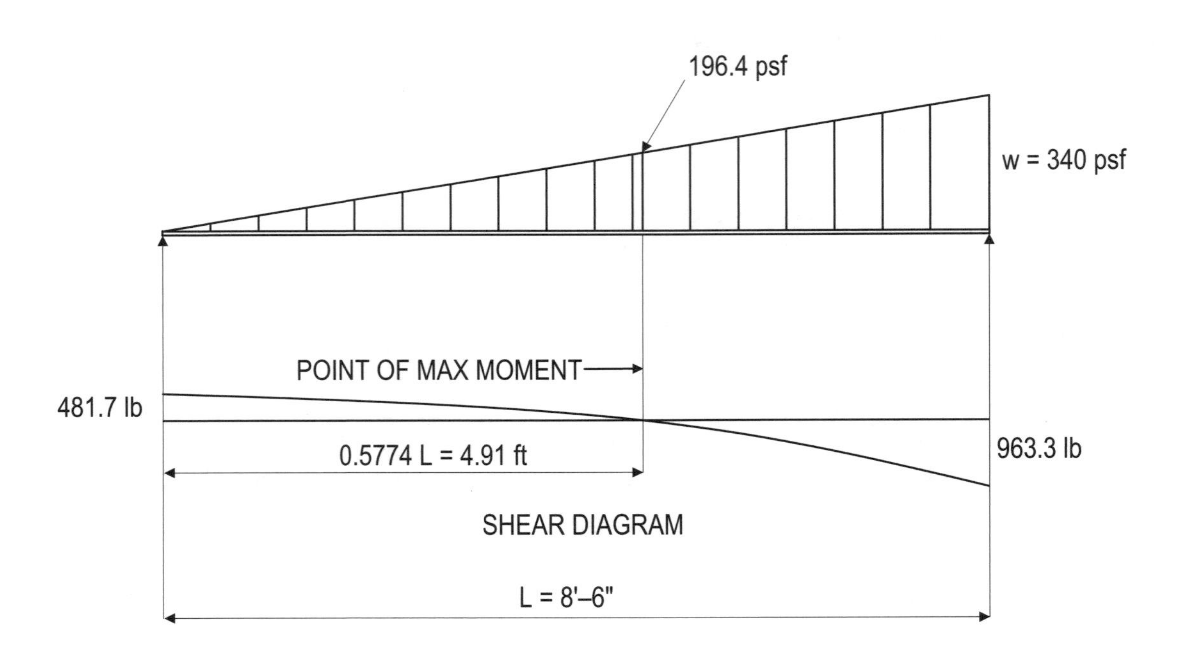

THE CORRECT ANSWER IS: (D)

111. Determine the effective flange width for composite design:

$b_{eff} = (1/4)$ span $= (1/4) \times 40 \times 12$ inches = 120 inches

or

$b_{eff} = 2\,(6 \times t_{slab})$ + flange width = $2\,(6 \times 7.5$ inches$) + 24$ inches = 114 inches ← controls

or

b_{eff} = c. to c. of girders = 10'-4" = 124 inches

The smaller value controls, therefore b_{eff} = 114 inches.

Determine the reinforcement area required for bending.

Note: the same "most nearly" answer is obtained for b_{eff} = 96 inches to 124 inches

Assume a rectangular section with b = 114 inches and d = 42 inches, then check a < slab thickness.

$M_u/\phi\, bd^2 = 2{,}000 \times 12{,}000$ lb-in/[0.90 × 114 inches × (42 in)^2] = 133 psi

ρ = (5,000 psi/60,000 psi) [1 – (1 – 4 × 0.59 × 133 psi/5,000 psi)$^{1/2}$]/(2 × 0.59)

= 0.00224 < ρ_{max} = 0.0267 OK

A_s = 0.00224 × 114 inches × 42 inches = 10.7 in^2, which is most nearly 11 in^2

Check stress block thickness:

a = $A_s f_y/0.85\, f_c b$ = 10.7 in^2 × 60,000 psi/(0.85 × 5,000 psi × 114 inches)

= 1.33 inches < 7.5 inches ← OK

THE CORRECT ANSWER IS: (B)

112. Determine the nominal shear strength of the concrete:

Note that since the location is not specified, the value of the term ($V_u d/M_u$) cannot be uniquely calculated; therefore, use the formula that is not a function of ($V_u d/M_u$).

b_w = 24 inches

$$V_c = 2 \times \sqrt{f'_c}\, b_w d = 2 \times \sqrt{5{,}000 \text{ psi}} \times 24 \text{ inches} \times 42 \text{ inches}/1{,}000$$

= 145 kips

THE CORRECT ANSWER IS: (D)

CIVIL MORNING SAMPLE SOLUTIONS

113. The solution to this problem is primarily based upon the *Highway Capacity Manual*, 1997, 3rd edition.

Based on Table 3-1 on page 3-11, the maximum service flow rate per lane for Level of Service D and free-flow speed of 70 mph is 2,048 pcphpl.

THE CORRECT ANSWER IS: (C)

114. Cost Buy:Cost Quarry

$$7(50{,}000)(P/A)^{10\%}_{10\ yr} = x + 5.5\,(50{,}000)(P/A)^{10\%}_{10\ yr}$$

$$x = 1.5\,(50{,}000)(P/A)^{10\%}_{10\ yr}$$

$$= \$460{,}900$$

If compound interest tables to 0.0001 are used, $(P/A)^{10\%}_{10\ yr} = 6.1446$ and answer = \$460,845.

THE CORRECT ANSWER IS: (C)

115. Approach:

a. Find the distance, T.
b. Add to P.C. station.

a. $T = R\ \text{Tan}\,\frac{\Delta}{2} = 2{,}550'\ \text{Tan}\,\frac{78^\circ 35'30''}{2} = 2086.84'$

b. 2,086.84 + Sta. 12 + 56.00 = Sta. 33 + 42.84

THE CORRECT ANSWER IS: (B)

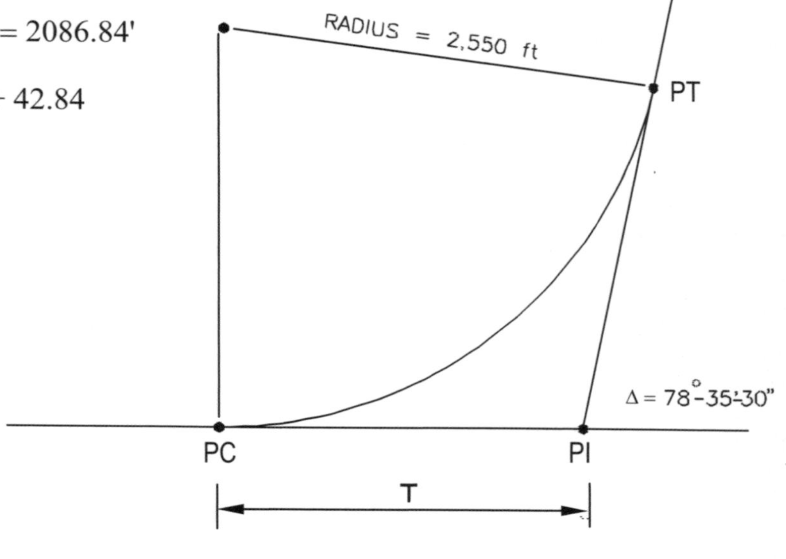

CIVIL MORNING SAMPLE SOLUTIONS

116. Stations on vertical curves are based on horizontal distances. Therefore, the required horizontal distance can be computed as follows:

Horizontal distance = (76 + 00.00) – (42 + 00.00) = 34 + 00.00
= 3,400 feet

THE CORRECT ANSWER IS: (B)

117. Determine the Hazen-Williams coefficient for new ductile iron pipe. The best estimate is 130.

THE CORRECT ANSWER IS: (B)

118. Determine which statement will reduce the tendency for pump cavitation.

II. Lowering the pump elevation
III. Increasing the suction diameter

THE CORRECT ANSWER IS: (A)

119. Natural grassland in an undeveloped condition is less impervious than 1/4-acre lots; therefore runoff will increase. Due to the increase in impervious area, the time of concentration for the site will decrease.

THE CORRECT ANSWER IS: (A)

120. Add calcium and magnesium concentrations in $CaCO_3$ equivalents.

Calcium = (51 mg/L) (50 mg/meq) / (20 mg/meq) = 127 mg/L as $CaCO_3$

Magnesium = (12 mg/L) (50 mg/meq) / (12.1 mg/meq) = 50 mg/L as $CaCO_3$

Hardness = (127 mg/L) + (50 mg/L) = 177 mg/L as $CaCO_3$

THE CORRECT ANSWER IS: (D)

CIVIL DEPTH

AFTERNOON SAMPLE SOLUTIONS

ENVIRONMENTAL

AFTERNOON SAMPLE SOLUTIONS

CORRECT ANSWERS TO THE ENVIRONMENTAL AFTERNOON SAMPLE QUESTIONS

Detailed solutions for each question begin on the next page.

501	C
502	B
503	B
504	C
505	B
506	C
507	A
508	C
509	A
510	B
511	D
512	D
513	D
514	C
515	D
516	C
517	C
518	C
519	C
520	A

501. Assuming specific gravity of sludge is 1.00
Sludge quantity and volume to be dewatered.

Primary Sludge

lb/d = (200 mg/L – 80 mg/L)(8.34)(20 MGD) = 20,016 lb/d

$$\text{gpd}\ \frac{20{,}016\ \text{lb/d}}{(8.34\ \text{lb/gal})(0.035)} = 68{,}571\ \text{gpd}$$

Secondary Sludge (was)

$$\text{gpd}\ \frac{10{,}000\ \text{lb/d}}{(8.34\ \text{lb/gal})(7{,}500\ \text{mg/L})} = 0.1599\ \text{MGD} = 159{,}872\ \text{gpd}$$

Before Thickening

Total quantity = 20,016 + 10,000 = 30,016 lb/d

Total volume = 68,571 + 159,872 = 228,443 gpd

After Thickening

Total quantity = 30,016 lb/d

$$\text{Total volume} = \frac{30{,}016}{(8.34\ \text{lb/gal})(0.045\ \text{percentage})} = 79{,}979\ \text{gpd}$$

$$\therefore\ \%\ \text{Reduction in sludge volume} = \frac{228.443 - 79.979}{228.443} = 65\%$$

THE CORRECT ANSWER IS: (C)

502. Determine the volume for a high-rate anaerobic digester.

The requisite digester volume must be evaluated separately for the solids and hydraulic loading criteria stated in the question.

Volatile Solids Loading:

$$\text{Maximum loading} = 100\,\frac{\text{lb VSS}}{1{,}000\ \text{ft}^3\text{- day}}$$

$$\text{Input VSS} = (10{,}000\ \text{lb SS/day})\left(0.70\,\frac{\text{lb VSS}}{\text{lb SS}}\right) = 7{,}000\ \text{lb VSS/day}$$

$$\text{Minimum volume} = \frac{7{,}000\ \text{lb VSS/day}}{100\,\dfrac{\text{lb VSS}}{1{,}000\ \text{ft}^3\text{- day}}} = 70{,}000\ \text{ft}^3 \quad \leftarrow \text{controls}$$

Hydraulic Loading:

Minimum hydraulic residence time = 15 days

Input flow rate = 15,000 gpd

$$\text{Minimum volume} = \frac{(15{,}000\ \text{gpd})(15\ \text{days})}{7.48\ \text{gal/ft}^3} = 30{,}080\ \text{ft}^3$$

The larger volume is required to satisfy the solids loading constraint.

Select a digester volume = 70,000 ft^3

Determine the pounds of digested sludge solids produced and the volume of the digested sludge leaving the first-stage digester.

The mass (dry weight) of output sludge would be less than the mass of input sludge to reflect the volatile suspended solids destruction:

output mass = input mass – volatile solids destruction

$$= 10{,}000\,\frac{\text{lb SS}}{\text{day}} - \left(10{,}000\,\frac{\text{lb SS}}{\text{day}}\right)\left(0.70\,\frac{\text{lb VSS}}{\text{lb SS}}\right)(0.60) = 5{,}800\ \text{lb VSS/day}$$

$$\therefore \text{Total solids leaving per unit volume of the digester} = \frac{5{,}800}{70{,}000} = 0.083\ \text{lb/}\left(\text{ft}^3\text{- day}\right)$$

THE CORRECT ANSWER IS: (B)

503. The raw wastewater contains 275 mg/L of BOD_5, and the primary clarifier is reported to remove 30% of the BOD_5. The concentration of BOD_5 in the primary clarifier effluent is calculated below.

$$BOD_5 = 275 \text{ mg/L } (1 - 0.30) = 192.5 \text{ mg } BOD_5/L$$

The treatment facility must achieve 95% removal of BOD_5. The raw wastewater contains 275 mg/L of BOD_5. The plant effluent BOD_5 concentration is calculated below:

$$BOD_5 = 275 \text{ mg/L } (1 - 0.95) = 13.75 \text{ mg } BOD_5/L$$

The sludge production may be determined with the following equation (Metcalf & Eddy. *Wastewater Engineering*, 3rd edition, McGraw-Hill, New York, NY, 1991, p. 377, 391):

$$P_x = \frac{Y\max}{1 + k_d \text{ SRT}} Q\left(S_o - S_e\right)$$

in which
- P_x = sludge production (dry mass/time)
- Y_{max} = maximum yield = 0.5 mg VSS/mg BOD_5
- k_d = endogenous decay coefficient, = 0.1 per day
- SRT = sludge age = 10 days
- Q = wastewater flow rate = 2.0 MGD
- s_o = BOD_5 in influent to activated sludge = 192.5 mg/L, and
- s_e = BOD_5 in the plant effluent = 13.75 mg/L

thus,

$$P_x = \frac{\left(0.5 \dfrac{\text{mg VSS}}{\text{mg BOD}_5}\right)}{1 + \left(0.1 \text{ day}^{-1}\right) 10 \text{ days}} \left(2 \text{ MGD}\right)\left(192.5 \text{ mg/L} - 13.75 \text{ mg/L}\right)\left(8.34 \frac{\text{lb/MGAL}}{\text{mg/L}}\right)$$

$$P_x = 745 \text{ lb/day}$$

The total inventory of biomass in the activated sludge process is equal to the product of the sludge production rate and the sludge age. The inventory is also equal to the product of the aeration basin volume (V) and the mixed liquor volatile suspended solids concentration (X).

$$V X = P_x \text{ SRT}$$

If the MLVSS is known, the volume of the aeration basin is calculated as follows:

$$V = \frac{P_x \text{ SRT}}{X} = \frac{\left(745 \text{ lb/day}\right)\left(10 \text{ days}\right)\left(10^6 \dfrac{\text{gal}}{\text{MGAL}}\right)}{\left(3{,}000 \text{ mg/L}\right)\left(8.34 \dfrac{\text{lb/MGAL}}{\text{mg/L}}\right)} = 297{,}800 \text{ gal}$$

THE CORRECT ANSWER IS: (B)

504. Solids in primary sludge = (250 × 2 × 8.34) 50% = 2,085 lb/day

$$\text{Specific gravity of primary sludge} = \frac{100}{\frac{4}{1.32} + \frac{96}{1.0}} = 1.01$$

$$\therefore \text{Volume of primary sludge} = \frac{2{,}085}{0.04 \times 1.01 \times 8.34} = 6{,}200 \text{ gpd}$$

Let secondary waste sludge flow = Q_w

$$X(Q+Q_r) = (Q_r + Q_w)X_r + (Q - Q_w)X_e$$

$$[2{,}500 \times (2 + 0.8)] = [(0.8 + Q_w) \times 8{,}000] + [(2 - Q_w) \times 25]$$

$$\therefore Q_w = 69{,}000 \text{ gpd}$$

THE CORRECT ANSWER IS: (C)

505. Determine, by calculating the flows involved, whether the existing sewer line can carry the wastewater flow from the new residential subdivision. Neglect inflow and infiltration. Reference all factors used.

Given: 5,775 people at 110 gpcd
36-inch RCP, n = 0.017
slope = 0.002 ft/ft
flow = 7 MGD maximum during day

(1) Calculate wastewater flow from subdivision

5,775 people × 110 gpcd = 635,250 gpd = 0.635 MGD = 0.982 cfs

(2) Calculate peak flow

0.63525 MGD × 3.8 = 2.414 MGD = 3.73 cfs

(3) Calculate potential peak flow for existing sewer

existing flow = 7 MGD

$$\text{existing flow} = 7\ \text{MGD} \times 1{,}000{,}000\ \frac{\text{ft}^3}{7.48\ \text{gal}} \times \frac{\text{day}}{86{,}400\ \text{sec}} = 10.83\ \text{cfs}$$

Peak flow after connection = 2.4 MGD + 7 MGD = 9.4 MGD

$$\text{Using Manning's Equation, full flow} = \frac{0.463}{0.017}(3)^{8/3}(0.002)^{1/2} = 14.71\ \text{MGD}$$

∴ Ratio Q/Q_f = 9.4/14.71 = 0.64

∴ d/D = 0.575 (Ref. Chow, *Open-Channel Hydraulics*, p. 135, Figure 6-5)

∴ d = 0.575 × 36 = 20.7 inches

THE CORRECT ANSWER IS: (B)

506. The BOD_5 in the river/effluent mixture after discharge is determined by a steady-state mass balance:

$$BOD_5 = \frac{Qe\ BOD_{5e} + Qr\ BOD_{5r}}{Qe + Qr}$$

in which BOD_5 = BOD_5 from point of discharge
BOD_{5e} = BOD_5 in effluent = 20 mg/L
BOD_{5r} = BOD_5 in river upstream from discharge = 4.0 mg/L

thus

$$BOD_5 = \frac{(6.19)(20) + (18)(4.0)}{(6.19 + 18)} = 8.09 \text{ mg/L}$$

The 5-day BOD is converted to the ultimate BOD (BOD_u) with the following equation (Metcalf and Eddy, 1991, p. 74):

$$BOD_5 = BOD_u\,[1 - \exp(-5\,k_1)]$$

in which k_1 = BOD reaction rate constant = 0.23 per day

thus

$$BOD_u = \frac{BOD_5}{[1 - \exp(-k_1 t)]} = \frac{8.09}{[1 - \exp(-(5)(0.23))]} = 11.84 \text{ mg/L}$$

THE CORRECT ANSWER IS: (C)

507. THE CORRECT ANSWER IS: (A)

508. For the specified conditions, the temperature at the point of discharge is 20°C. The corresponding DO_{sat} = 9.08 mg/L (Metcalf and Eddy, 1991, p. 1258). Thus, the initial deficit D_{init} is:

$$D_{init} = DO_{sat} - DO = 9.08 - 6.0 = 3.08 \text{ mg/L}$$

Other parameters are defined as follows:

k_1 = 0.23 per day
k_2 = 0.40 per day
BOD_u = 10 mg/L
velocity = 1.0 fps = 16.36 miles per day

The time to achieve the maximum dissolved oxygen deficit is determined with the following equation (Metcalf and Eddy. 1991, p. 1217).

$$t_{max} = \frac{1}{k_2 - k_1} \ln\left[\frac{k_2}{k_1}\left(1 - \frac{D_{init}\,(k_2 - k_1)}{k_1\ BOD_u}\right)\right]$$

in which t_{max} = time to achieve the maximum deficit.

For the stated conditions:

$$t_{max} = \frac{1}{0.4 - 0.23} \ln\left[\frac{0.4}{0.23}\left(1 - \frac{3.08\,(0.4 - 0.23)}{(0.23)\,(10)}\right)\right] = 1.74 \text{ days}$$

The corresponding distance is obtained by multiplication by the velocity:

Distance = velocity t_{max} = (16.36) (1.74) = 28.4 miles

THE CORRECT ANSWER IS: (C)

509. THE CORRECT ANSWER IS: (A)

510. $k_{27} = k_{20}\ 1.047^{T-20}$

$k_{27} = 0.23\ 1.047^{27-20} = 0.32$

$$BOD_L = \frac{BOD_T}{\left(1 - e^{-kT}\right)} = \frac{100}{\left(1 - e^{-0.32(7)}\right)} = 112 \text{ mg/L}$$

$$BOD_T = BOD_L\left(1 - e^{-kt}\right) = 112\left(1 - e^{-0.23(5)}\right) = 76.5 \text{ mg/L}$$

THE CORRECT ANSWER IS: (B)

511. Volume of raw waste = $\dfrac{6{,}000 \times 2{,}000}{300} = 40{,}000 \text{ yd}^3$

Volume of compacted waste = $40{,}000 \times 0.30 = 12{,}000 \text{ yd}^3$

$\therefore$ Volume of soil cover = $\dfrac{12{,}000}{10} = 1{,}200 \text{ yd}^3$

THE CORRECT ANSWER IS: (D)

512. From the table in **Question 512**, Solid Waste Characteristics, stream composition and average moisture % by weight are given. Consider 100 lb of the average solid waste stream. Determine the pounds of moisture in each component (% by weight of component = lb of component times the moisture % by weight). Sum the total weight of moisture and divide by the 100 lb of the average solid waste. Calculations are shown in the table below.

Sum of the moisture = 21.42 lb; 21.42 lb moisture/100 lb waste = 21.42% moisture by weight.

Column 1	Column 2	Column 3
Component %	**Moisture Content**	**Column 1 × Column 2**
8	0.65	5.2
39	0.07	2.73
6	0.06	0.36
5	0.025	0.125
2.5	0.105	0.2625
1	0.025	0.025
0.3	0.1	0.03
20.5	0.55	11.275
3	0.275	0.825
6	0.025	0.15
5.7	0.03	0.171
3	0.09	0.27

THE CORRECT ANSWER IS: (D)

ENVIRONMENTAL AFTERNOON SAMPLE SOLUTIONS

513. Little Creek Flow Upstream of Treatment Facility = 1 cfs
Treated groundwater flow = 100 gpm

Upstream Contaminants:

Tetrachloroethylene	0 μg/L
Toluene	0 μg/L
Chromium VI	30 μg/L
Lead	10 μg/L

Groundwater Contaminants:

Tetrachloroethylene	165 μg/L
Toluene	7,300 μg/L
Chromium VI	3,800 μg/L
Lead	200 μg/L

EPA Downstream Limits:

Tetrachloroethylene	1 μg/L
Toluene	875 μg/L
Chromium VI	50 μg/L
Lead	120 μg/L

Calculate allowable effluent limits for treated groundwater.

Set up mass balance:

$$\text{Downstream limit} = \frac{QoCo + Q1C1}{Qo + Q1}$$

$$\text{Rearrange to find Cl} = \frac{C2\left(Qo + Q1\right) - QoCo}{Q1}$$

Qo = 1 cfs Q1 = 100 gpm = 0.222 cfs

Tetrachloroethylene

$$\text{Cl} = \frac{1\ \mu g/L\left(1 + 0.222\right) - 1\left(0\right)}{0.222} = 5.5\ \mu g/L$$

$$\text{Toluene} = \frac{875\ \mu g/L\left(1 + 0.222\right) - 1\left(0\right)}{0.222} = 4{,}816\ \mu g/L$$

$$\text{Chromium VI} = \frac{50\ \mu g/L\left(1 + 0.222\right) - 1\left(30\right)}{0.222} = 140\ \mu g/L$$

513 (Continued)

$$\text{Lead} = \frac{120\ \mu g/L\,(1+0.222)-1\,(10)}{0.222} = 615\ \mu g/L$$

Percent Removals:

$$\text{Tetrachloroethylene:}\ \frac{165-5.5}{165} = 96.6\,\%$$

$$\text{Toluene:}\ \frac{7{,}300-4{,}816}{7{,}300} = 34.0\%$$

$$\text{Chromium VI:}\ \frac{3{,}800-140}{3{,}800} = 96.3\%$$

$$\text{Lead:}\ \frac{200-615}{200} = -207\%\ \text{No removal necessary}$$

THE CORRECT ANSWER IS: (D)

514. Constant Head Permeability Test:

V = volume = 10 in^3

L = sample length = 6 inches

A = area of sample = 1/4 × π (3 in)^2 = 7.07 in^2

H = head loss = (30 inches – 10 inches) = 20 inches

t = time of flow = 2 minutes

$$k = \frac{VL}{AHt} = \frac{(10\ in^3) \times (6\ in)}{(7.07\ in^2) \times (20\ in) \times (2\ min)} = 0.21\ in/min$$

THE CORRECT ANSWER IS: (C)

515. Falling Head Permeability Test:

a = cross-sectional area of burette = 1/4 × π (1 in)^2 = 0.785 in^2

L = length of sample = 6 inches

A = area of sample = 1/4 × π (3 in)^2 = 7.07 in^2

t = time of flow = 20 minutes

h_1 = initial head in burette = 20 inches

h_2 = final head in burette = 10 inches

$$k = \frac{aL}{At} \ln\left(\frac{h_1}{h_2}\right) = \frac{(0.785\ in^2) \times (6\ in)}{(7.07\ in^2) \times (20\ min)} \times \ln\left(\frac{20\ in}{10\ in}\right) = 0.023\ in/min$$

THE CORRECT ANSWER IS: (D)

516. Determine the pump efficiency at the operating point in the system for an 8-inch pipe. Use Hazen-Williams equation and calculate the total dynamic head for a nominal 8-inch pipe diameter.

$Q = V\,A = 1.318\ C\ A\ R^{0.63}\ S^{0.54}$

$C = 100$

$A = (1/4)\ \pi D^2 = (1/4)\ (3.1416)\ (0.667)^2 = 0.3490\ ft^2$

$R = D/4 = (0.667\ /\ 4) = 0.1667\ ft$

$$S^{0.54} = \frac{Q}{1.318\ C\ A\ R^{0.63}} = \frac{Q}{1.318\,(100)\,(0.3490)\,(0.1667)^{0.63}} = 0.06721\,Q$$

$S = 0.006740\ Q^{1.8519}$

$h_f = (3{,}000\ ft)\ (0.006740)\ Q^{1.8519}$

$TDH = \Delta H + h_f + V^2/2\,g$ and $\Delta H = 560\ ft - 530\ ft = 30\ ft$

Q (gpm)	Q (cfs)	h_f (ft)	V = Q/A (fps)	$V^2/2\,g$ (ft)	ΔH (ft)	TDH (ft)
400	0.891	16.337	2.554	0.101	30.00	46.4
600	1.337	34.617	3.830	0.228	30.00	64.8
800	1.783	58.974	5.107	0.405	30.00	89.4

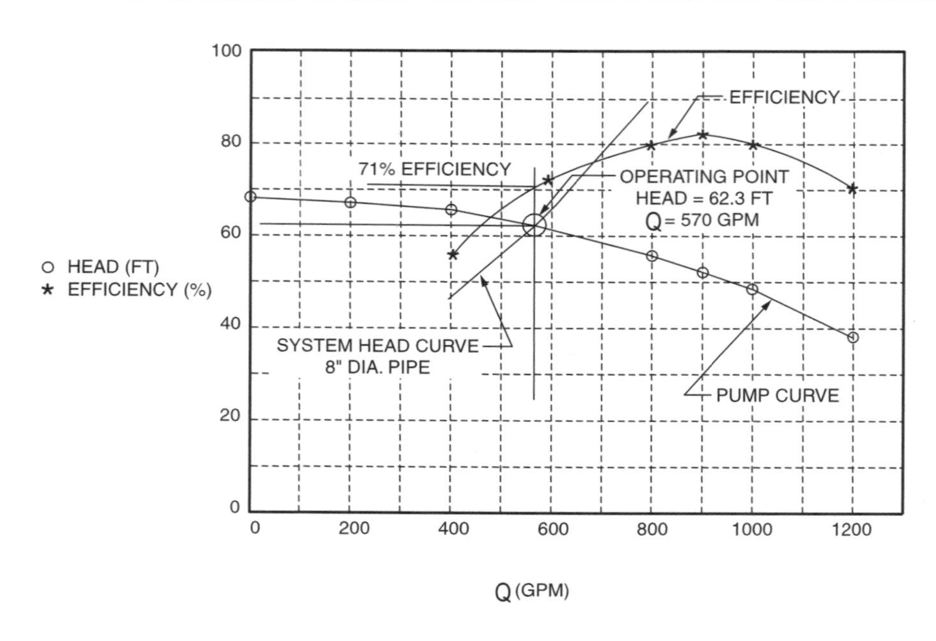

Plot the system head curve (TDH). Where the system head curve crosses the pump curve, project that point up to the efficiency curve.

Efficiency = 71%

THE CORRECT ANSWER IS: (C)

517. Determine the water velocity for the trapezoidal ditch. For a depth = 2.0 ft.

$A = (2.0 + 3 \times 2.0)(2.0) = 16.0\ ft^2$

$$R = \frac{(b + zy)\,y}{b + 2y\sqrt{1+z^2}} = \frac{(2.0 + 3 \times 2.0)(2.0)}{2.0 + 2 \times 2.0\sqrt{1+3^2}} = 1.092\ ft$$

$$V = \frac{1.486}{n} R^{2/3} s^{1/2} = \frac{1.486}{0.02} \times (1.092)^{2/3} (0.005)^{1/2} = 5.6\ ft/s$$

THE CORRECT ANSWER IS: (C)

518. Determine the SCS Runoff Curve Number for the entire area. For soil Group B, with good vegetative cover in urban, fully developed open space (parks, lawns), the appropriate SCS curve number (Gupta, *Hydrology and Hydraulic Systems*, Prentice Hall, 1989, p. 101) CN = 61.

THE CORRECT ANSWER IS: (C)

519. Surface Water Treatment Rule for disinfection of *Giardia*; 3 log inactivation required. 2.5 log inactivation allowed for treatment prior to disinfection. Therefore, 3 – 2.5 = 0.5 log inactivation required by disinfection.

From the figure, at a peak hourly flow rate of 5 MGD, t_{10} = 50 min contact time.

CT = 23 mg/L•min; CT = concentration × time

Ref: Davis & Cornwell, *Environmental Engineering*, McGraw Hill, 1998, p. 244, Table 3-20.

C = CT/T = (23 mg/L•min)/50 min = 0.46 mg/L residual chlorine concentration needed.

THE CORRECT ANSWER IS: (C)

520. Nitrate > 10 mg/L as NO_3 -N

Turbidity > 1–5 NTU

Coliforms > 1 organism/100 ml

THE CORRECT ANSWER IS: (A)

GEOTECHNICAL

AFTERNOON SAMPLE SOLUTIONS

CORRECT ANSWERS TO THE GEOTECHNICAL AFTERNOON SAMPLE QUESTIONS

Detailed solutions for each question begin on the next page.

501	C
502	C
503	B
504	C
505	C
506	D
507	C
508	A
509	B
510	B
511	D
512	C
513	D
514	D
515	A
516	B
517	D
518	C
519	B
520	B

GEOTECHNICAL AFTERNOON SAMPLE SOLUTIONS

501. Sample A has the following properties:

% passing the #4 sieve	77%
% fines passing the #200 sieve	18%
% retained on the #200 sieve	100% – 18% = 82%
Liquid Limit, LL	32%
Plastic Limit, PL	25%
Plasticity Index, PI = LL – PL	7%

Based on 77% passing the #4 sieve and 82% retained on the #200 sieve, the soil is classified as a sand, either SM or SC. Based on the fines having LL = 32 and PI = 7, the fines would be classified as ML, nonplastic. Therefore, according to the Unified Soil Classification System, Sample A is classified as SM.

THE CORRECT ANSWER IS: (C)

502. Sample B has the following properties:

% passing the #200 sieve	55%
Liquid Limit, LL	52%
Plastic Limit, PL	32%
Plasticity Index, LL – PL	20%
LL – 30 = 22, >20	

Based on 55% passing the #200 sieve, the soil may be classified as A-4, A-5, A-6, A-7-5, or A-7-6. Based on LL = 52 and PI = 20, the soil cannot be classified as A-4, A-5, nor A-6. Since PI < LL – 30, according to the AASHTO Classification System, Sample B is classified as A-7-5.

THE CORRECT ANSWER IS: (C)

503. Compute the volume of Sample A:

$$V = \frac{\pi \times d^2 \times h}{4} = \frac{\pi \times (3\text{ in})^2 \times (6\text{ in})}{4} = 42.4\text{ in}^3 = 0.0245\text{ ft}^3$$

Compute the volume of solids:

$$V_s = \frac{W_s}{G_s \times \gamma_w} = \frac{(2.54\text{ lb})}{(2.65) \times (62.4\text{ pcf})} = 0.0154\text{ ft}^3$$

Compute volume of voids:

$$V_v = V - V_s = 0.0245\text{ ft}^3 - 0.0154\text{ ft}^3 = 0.0091\text{ ft}^3$$

Compute void ratio:

$$e = \frac{V_v}{V_s} = \frac{0.0091\text{ ft}^3}{0.0154\text{ ft}^3} = 0.59$$

Compute moisture content:

$$w = \frac{W_w}{W_s} = \frac{(2.95\text{ lb} - 2.54\text{ lb})}{(2.54\text{ lb})} = 0.161\text{ or }16.1\%$$

Compute degree of saturation:

$$S = \frac{G_s \times w}{e} = \frac{(2.65) \times (0.161)}{(0.59)} = 0.72\text{ or }72\%$$

THE CORRECT ANSWER IS: (B)

504. Sample F—Triaxial Test

$\sigma_1 = 33.5$ psi, $\sigma_3 = 16.4$ psi, $u = 10.0$ psi

$\sigma_1' = \sigma_1 - u = 23.5$ psi, $\sigma_3' = \sigma_3 - u = 6.4$ psi

$$p = \frac{\sigma_1' + \sigma_3'}{2} = \frac{23.5 \text{ psi} + 6.4 \text{ psi}}{2} = 14.95 \text{ psi}$$

$$q = \frac{\sigma_1' - \sigma_3'}{2} = \frac{23.5 \text{ psi} - 6.4 \text{ psi}}{2} = 8.55 \text{ psi}$$

$$\alpha' = \arctan\left(\frac{q}{p}\right) = \arctan\left(\frac{8.55 \text{ psi}}{14.95 \text{ psi}}\right) = 29.8°$$

$$\phi' = \arcsin(\tan\alpha) = \arcsin(\tan 29.8°) = 34.9°$$

THE CORRECT ANSWER IS: (C)

505. The quantity of seepage loss under the dam may be computed using the flow net as follows:

$$Q_1 = kHL\left(\frac{N_f}{N_d}\right) \text{ where:}$$

N_f = number of flow channels

N_d = number of equipotential drops

$$Q_1 = (0.003 \text{ fps})(12 \text{ ft})(120 \text{ ft})\left(\frac{6}{15}\right)$$

$$Q_1 = 1.73 \text{ cfs}$$

THE CORRECT ANSWER IS: (C)

506. The pressure head at Point A is determined as follows:

Total head loss, headwater to tail water = 12.0 feet.

Number equipotential drops = 15, or 0.80 foot/drop.

Number equipotential drops, headwater to Point A = 9.5. Therefore, the total head loss from the headwater to Point A equals 9.5 drops × 0.80 foot/drop = 7.6 feet.

Total head at Point A = (12.0 feet – 7.6 feet) or 4.4 feet.

The pressure head at Point A equals the total head minus the elevation head. Note that the elevation head is measured with respect to the tail water. Therefore, the pressure head at Point A equals [4.4 feet – (–11.0 feet)] = 15.4 feet.

THE CORRECT ANSWER IS: (D)

507. Determine the primary consolidation settlement at the middle of the clay layer.

$$\Delta H = \frac{H_o}{(1+e_o)} C_c \log\left(\frac{\sigma'_o + \Delta\sigma}{\sigma'_o}\right) = \frac{10.0}{(1+1.0)}(0.5)\log\left(\frac{1{,}000+200}{1{,}000}\right)$$

$\Delta H = 0.20$ ft = 2.4 inches

THE CORRECT ANSWER IS: (C)

508. Determine the time required to achieve 50% primary consolidation.

For the double drainage condition, $H_{DR} = 1/2\ H_o = 1/2\ (10.0) = 5.0$ ft = 60 inches

From standard reference table for U = 50%, $T_v = 0.197$

$T = T_v H^2_{DR}/C_v = (0.197)(60^2)/1.6 \times 10^{-4} = 4.43 \times 10^6$ sec

$= (4.43 \times 10^6$ sec) (min/60 sec) (hr/60 min) (day/24 hr) (mo/30 day)

= 1.7 mo, which is most nearly 2 months.

THE CORRECT ANSWER IS: (A)

509. Pile end bearing capacity by Terzaghi equation.

$Q_{PB} = \sigma'_v \ Nq \ \text{area}$

$\sigma'_v = 3(100 \text{ pcf}) + 4(100 - 62.4) + 6(120 - 62.4) = 796 \text{ psf}$

$N_q = 260 \text{ (given)} \quad \text{Area} = \dfrac{\pi D^2}{4} = \dfrac{\pi(1)}{4} = 0.79 \text{ ft}^2$

$Q_{PB} = 796(260)(0.79) = 163{,}000 \text{ lb}$

$Q_{PB} = 82 \text{ tons}$

THE CORRECT ANSWER IS: (B)

510. Ultimate skin friction using Mohr-Coulomb failure criteria

$Q_{SF} = \sum \sigma'_v \ K \tan\delta \ \text{Area}_{surf}$

$\text{Area} = \pi D \Delta L = \pi(1)\Delta L$

Depth (ft)	σ'_v	K_o	Tan δ	Area (ft)
3	3 (100) = 300 psf	0.5	0.325	9.42
7	300 + 4 (100 – 62.4) = 450	0.5	0.325	12.57
13	450 + 6 (120 – 62.4) = 796 psf	1.0	0.466	18.85

$$Q_{SF} = \frac{300}{2}(0.5)(0.325)(9.42) + \left(\frac{300 + 450}{2}\right)(0.5)(0.325)(12.57) + \left(\frac{450 + 796}{2}\right)(1.0)(0.466)(18.85)$$

$Q_{SF} = 3.3 \text{ tons}$

THE CORRECT ANSWER IS: (B)

511. Determine the active lateral earth resultant on the lowest facing unit.
$P_A = (q_o + \gamma h) K_A$ where h = depth below the top of the embankment.

At EL. 3 feet, h = 12 – 3 = 9 feet

P_A at 9-foot depth = {(200 + (115) (9.0)} (0.30) = 371 psf

At EL. 0 foot, h = 12 – 0 = 12 feet

P_A at 12-foot depth = {(200 + (115) (12.0)} (0.30) = 474 psf

R = (1/2) (P_A at 9-foot depth + P_A at 12-foot depth) (area of facing unit)

= (1/2) (371 + 474) (3.0 × 3.0) = 3,800 lb

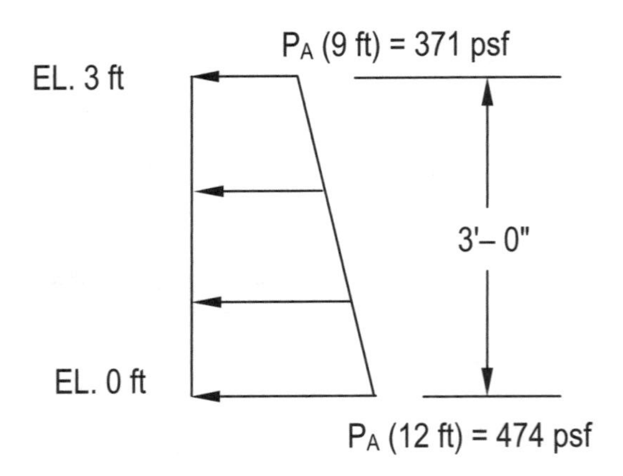

THE CORRECT ANSWER IS: (D)

512. Determine the length of the reinforcing strip required beyond the failure surface.

σ'_a = $\sigma'_v K_A = (\sigma'_v) (0.30)$

w = width of strip = 12 inches = 1.0 foot

ΔL = (F.S) (σ'_a / σ'_v) (area of facing unit)/(2 w tan 22°)

= (3.0) (0.30) (3.0 × 3.0)/(2 × 1.0 × 0.404) = 10 feet

THE CORRECT ANSWER IS: (C)

513. The cyclic stress ratio = 0.29 and σ'_v = 1.1 tsf.

$$0.29 = \frac{\tau}{\sigma'_v} = \frac{\tau}{1.1 \text{ tsf}}$$

$$\tau = (0.29)\,(1.1) = 0.319 \text{ tsf} = 638 \text{ psf}$$

The earthquake-induced shear stress (450 psf) is less than the shear stress (τ) to cause liquefaction (638 psf), thus the factor of safety is:

$$\text{Factor of safety} = \frac{\tau}{450 \text{ psf}} = \frac{638 \text{ psf}}{450 \text{ psf}} = 1.4$$

THE CORRECT ANSWER IS: (D)

514. (A) is incorrect because the "flowing artesian" effect applies to wells not aquifers, and in any event in the situation described, the piezometric surface of the deep aquifer is below the ground surface and not flowing out above the ground surface.

(B) is incorrect because the aquifer is confined by the upper clay layers within a piezometric level well above the bottom of the clay layer.

(C) obviously does not apply to a coarse sand and gravel aquifer.

THE CORRECT ANSWER IS: (D)

515. The commonly accepted method of design in the question setting would be to drain the shallow surficial aquifer to the nearby stream using gravity flow devices, and then to excavate down into the underlying clay to obtain both sand for cover materials and clay for liner and capping materials to maximize the site volume and economic return, and to take advantage of the upward hydraulic pressures to minimize the potential for leachate migration.

(B) would substantially add to the cost and hydraulic controls would still be necessary.

(C) slurry walls are extremely costly and hydraulic controls would still be necessary.

(D) a leak-proof liner does not exist.

THE CORRECT ANSWER IS: (A)

516. Find unit weight (pcf) for Trial Mix No. 1.

Given: Volume $= 2.27\ ft^3$

$Weight_{coarse} = 130.0$ lb Cement $= 50.0$ lb

$Weight_{fine} = 100.0$ lb Water $= 30.0$ lb

Then: Total weight = 130 + 100 + 50 + 30 = 310 lb

$$\therefore \text{unit weight} = \frac{\text{total weight}}{\text{volume}} = \frac{310}{2.27} = 136.56 \text{ pcf}$$

THE CORRECT ANSWER IS: (B)

517. Find percent volume of aggregate for Trial Mix No. 2.

Given: Final Volume $= 2.30\ ft^3$

$Weight_{coarse} = 129.0$ lb; moisture content = 2.5%; SSD = 0.5%; S.G. = 2.65

$Weight_{fine} = 101.0$ lb; moisture content = 6.5%; SSD = 1.5%; S.G. = 2.70

Find volumes:

$$Volume_{coarse} = \frac{129.0}{1+0.025} \times (1+0.005) \div 2.65 \times 62.4 = 0.76\ ft^3$$

$$Volume_{fine} = \frac{101.0}{1+0.065} \times (1+0.015) \div 2.70 \times 62.4 = 0.57\ ft^3$$

Total Volume Aggregates $= 1.33\ ft^3$

$$\therefore \text{Percent of volume} = \frac{1.33}{2.30} \times 100\% = 58\%$$

THE CORRECT ANSWER IS: (D)

518. Determine the maximum unfactored moment with tieback.

$Ph = 3/(10)^{1/2} \times 1.0\ k/ft^2 = 0.95\ k/ft^2$

$Th = 2/(5)^{1/2} \times 10$ kips $= 8.9$ kips

$M = (0.95\ k/ft^2 \times 5\ ft \times 18\ ft \times 12\ ft) - 8.9\ k \times 21\ ft = 839$ k-ft

THE CORRECT ANSWER IS: (C)

519. Determine the minimum bonded length of tieback.

Allowable frictional resistance = 10 psi/1.5 = 6.67 psi

Circumference of drilled hole = 6 inches × 3.14 = 18.8 inches

Min L = 10 kips × 1,000/6.67 psi × 18.8 inches × 12 = 6.6 ft

THE CORRECT ANSWER IS: (B)

520. Determine the minimum number of truckloads from the borrow pit.

The embankment construction requires 500,000 yd^3 of soil at:

$\gamma_{dry} = (0.90)\,(120.0) = 108.0$ pcf

The total weight of dry soil required is:

$W_{total} = (500{,}000\ yd^3)\,(27\ ft^3/yd^3)\,(108.0\ pcf) = 1.458 \times 10^9$ lb

The dry unit weight of soil in the truck is:

$\gamma_{dry} = G_s\,\gamma_w/(1 + e) = (2.65)\,(62.4)\,/(1 + 1.30) = 71.9$ psf

Each truck can carry a weight of:

$W_{truck} = (5.0\ yd^3)\,(27\ ft^3/yd^3)\,(71.9\ pcf) = 9{,}700$ lb/truck

Therefore, the minimum number of trucks required is:

$N = W_{total}/W_{truck} = 1.458 \times 10^9/9{,}700 = 150{,}000$ trucks

THE CORRECT ANSWER IS: (B)

STRUCTURAL

AFTERNOON SAMPLE SOLUTIONS

CORRECT ANSWERS TO THE STRUCTURAL AFTERNOON SAMPLE QUESTIONS

Detailed solutions for each question begin on the next page.

501	D
502	B
503	B
504	C
505	B
506	B
507	A
508	A
509	A
510	B
511	B
512	C
513	A
514	A
515	B
516	C
517	D
518	B
519	B
520	B

STRUCTURAL AFTERNOON SAMPLE SOLUTIONS

501.

$$W = \left[\frac{75}{1{,}000}\left(\frac{26^2}{2(23)}\right)(4(80)+2(100))\right] + \left[\frac{10}{1{,}000}\left(\frac{26^2}{2(23)}\right)2(40)\right] + \frac{[15(199)(99)]}{1{,}000} = 880 \text{ kips}$$

Alternate Solution

$$W = \left[\frac{75}{1{,}000}\left(\frac{23}{2}+3\right)(4(80)+2(100))\right] + \left[\frac{10}{1{,}000}\left(\frac{23}{2}+3\right)2(40)\right] + \frac{[15(199)(99)]}{1{,}000} = 873 \text{ kips}$$

THE CORRECT ANSWER IS: (D)

502. UBC Solution

Soil default $S_D, C_a = 0.36, C_v = 0.54$

$R = 4.5$ (Table 16N item 1, 2, a)

$I = 1.0$

$$V = \frac{0.54(1)w}{4.5(0.21)} = 0.57 \text{ w} \quad \text{or,}$$

$$V = \frac{2.5(0.36)(1)w}{4.5} = 0.20 \text{ w} \quad \text{but,}$$

$$V_{mu} = 0.11(0.36)(1)w = 0.04 \text{ w} \quad \text{so,}$$

$$V = 0.20 \text{ w}$$

SBC / NBC Solution

Soil default $S = 2.0 \quad I = 1.0$

$R = 4.5$ (Table 1610.33)

$$C_s = \frac{1.2(0.30)(2.0)}{4.5(0.21)^{2/3}}$$

$$C_s = 0.45 \quad \text{or,}$$

$$C_s = \frac{2.5(0.30)}{4.5} = 0.17$$

THE CORRECT ANSWER IS: (B)

503. As per AASHTO 3.24.3.1

$$M_{L-L} = \left\{\left(\frac{S+2}{32}\right)P_{20}\right\}(0.8) \quad \text{for continuity}$$

$$= \left(\frac{9.5+2}{32}\right)(16)(0.8) = 4.6 \text{ ft-kips}$$

THE CORRECT ANSWER IS: (B)

504.

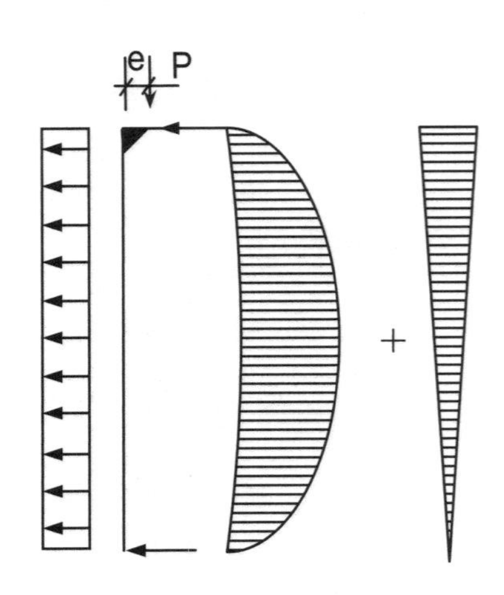

$P_{roof} = 12\ (15 + 40) = 660$ plf

$$e = \frac{7.625}{2} + 3.5 = 7.31 \text{ inches}$$

$M_{roof} = 660\ (7.31)/12 = 402$ lb-ft/ft

M_{wind} @ midheight $= 20\ (12)^2/8 = 360$ lb-ft/ft

$\Sigma M_{midheight} = 360 + 402/2 = 561$ lb-ft/ft

Alternative couple (shear) $\quad \dfrac{402}{12} = 33.5, \quad 20 \times \dfrac{12}{2} = 120$

Superimpose shears (top) = 33.5 + 120 = 153.5

$$V = 0 \ @ \ \frac{153.5}{20} = 7.67 \text{ ft from top}$$

$$\therefore \ M \ @ \ 7.67 \text{ ft} = 153.5 \times \frac{7.67}{2} = 588 \text{ ft-lb}$$

THE CORRECT ANSWER IS: (C)

505. $f'_m = 1{,}500$ psi net area compressive strength = 1,900 (Table 2, p. 5-10)

$$E_n = 2.2 - \left(\frac{100}{500}\right)(0.6) = 2.08 \times 10^6 \quad n = 29/2.08 = 13.94 \quad \text{say } 14$$

(This is linear interpolation of Table 5.5.2.3.)

ACI 530 states that members shall be designed by elastic analysis, using service loads (working stress design). Equations below are working stress equations. (See textbooks.)

$$p = 0.31 \Big/ 48\left(\frac{7.625}{2}\right) = 0.0017 \quad np = 0.0238$$

$$k = \sqrt{np^2 + 2np} - np = \sqrt{(0.0238)^2 + 2(0.0238)} - 0.00238 = 0.195$$

$j = 1 - k/3 = 0.935 \qquad F_b = 1.33\ (1/3)\ f'_m = 1.33\ (1/3)\ 1{,}500 = 667$ psi

$M_{max} = F_b bkjd^2/2\ (12) = 667\ (12)\ (0.195)\ (0.935)\ (7.625/2)^2\ /2\ (12)$

$M_{max} = 885$ lb-ft/ft

THE CORRECT ANSWER IS: (B)

506. $w = 400$ plf

$$m = \frac{wl^2}{8}$$

$$= \frac{(400)(192)^2}{8(1{,}000)} = 1{,}843 \text{ ft-kips}$$

$$\text{Chord Force} = \frac{m}{120} = \frac{1{,}843}{120} = 15.4 \text{ kips}$$

120'

192'

w = 400 plf

THE CORRECT ANSWER IS: (B)

507.

ASD

Sign Col.

$$\text{direct shear} = [4(8) + 25(1.0)(0.67)] \times 0.036 = 1.76$$

$$f = \frac{17.6}{13.7} = 0.13$$

$$\text{Torque} = 4(8)(0.036)15 = 17.28 \text{ ft-kips}$$

$$f = \frac{Mc}{I_{polar}} = \frac{17.28(12)(6)}{2(231)}$$

$$= 2.69 \text{ ksi}$$

$$f_{total} = 0.13 + 2.69 = 2.82 \text{ ksi}$$

LRFD

$$1.3\ f = 1.3(2.8) = 3.64$$

(Apply load factor for LRFD design)

THE CORRECT ANSWER IS: (A)

508. ASD option:

$f_{bx} = 20.0$ $P = 7.2 + 12.6 + 4.0 = 23.8$ kips

$F_{ex}' = 40.1$ ksi per Table 8 on p. 5-122, AISC-ASD, 9th edition

$F_{bx} = 24.0$ (incl. 1/3 increase for wind)

$D + L + W = 23.8$ kips and $A = 15.6\ in^2$

$f_a = 23.8/15.6\ in^2 = 1.53$ ksi $< 0.15\ F_a$ $\frac{k_x l_x}{r_x} = \frac{2.0(15)12}{5.89} = 61.1$

$F_a = 13.75$ ksi per AISC Table C.36 $\frac{k_y l_y}{r_y} = \frac{1.0(15)12}{1.92} = 93.8$

So AISC eq. H1–3 $\frac{1.53}{13.75 \times 1.33} + \frac{20}{24} = 0.084 + 0.833 = 0.917$

THE CORRECT ANSWER IS: (A)

508. LRFD option:

$P_u = (1.2 \times 7.2) + (1.6 \times 12.6) + (0.80 \times 4.0) = 32.0$ kips (eq. A4.3) ← Controls

or $= (1.2 \times 7.2) + (1.3 \times 4.0) + (0.5 \times 12.6) = 20.1$ kips (eq. A4.4)

$$\begin{bmatrix} \frac{k_x l_x}{r_x} = \frac{(2)(15 \times 12)}{5.89} = 61.1 \\ \frac{k_y l_y}{r_y} = \frac{(1)(15 \times 12)}{1.92} = 93.8 \quad \text{(Govern)} \end{bmatrix}$$

$\phi P_n = 300.5$ kips ($k_y l_y = 15$ feet) (chart on p. 3-21, AISC-LRFD, 2nd edition)

$$\frac{32.0}{300.5} = 0.106 < 0.2 \quad \text{use (H1-1b)}$$

$$\frac{P_u}{2\phi P_n} + \frac{M_{ux}}{\phi_b M_{ux}} = \frac{32}{(2)(300.5)} + \left(\frac{180}{235}\right) = 0.819$$

THE CORRECT ANSWER IS: (A)

509. When concrete is placed on a hot windy day, it is not permissible to add field water as needed to obtain the desired consistency and workability.

THE CORRECT ANSWER IS: (A)

510. $$\text{Impact} = \frac{50}{L+125} \qquad (3-1)$$

$$= \frac{50}{85+125} = 0.24$$

$$\text{Dist.} = \frac{5}{5.5} \qquad (3.23.1)$$

$$= \frac{8.5}{5.5} = 1.55$$

$$\text{Mom.} = (1.55)(1.24)\left(\frac{1{,}254.7}{2}\right) \leftarrow \text{Appendix A, AASHTO, 16th edition}$$

$$= 1{,}206 \text{ ft-kips}$$

THE CORRECT ANSWER IS: (B)

511. ASD option:

$V = 1.5/4 = 0.38$ kips/bolt

$M = 12 \times 1.5 = 18.0$ ft-kips

Diagonal dimension = $8 \times \sqrt{2} = 11.3$ inches

$$T_{max} = \frac{18 \times 12}{11.3} = 19.1 \text{ kips}$$

Try 1"ϕ bolt $A_D = 0.785$ in^2

$$f_v = \frac{0.38}{0.785} = 0.48 \text{ ksi}$$

$F_t = 26(1.33) - 1.8(0.48) = 33.72 > 1.33(20)$

$T_{allowable} = 20.0$ ksi $\times 0.785 \times 1.33 = 20.9$ ksi > 19.1 Use 1"ϕ bolt

THE CORRECT ANSWER IS: (B)

511. LRFD option:

$M_u = 1.3 \times 1.5 \times 12 = 23.4$ ft-kips

$V_u = 1.3 \times 1.5 = 1.95$ kips

Diagonal dimension = 11.3 kips

Try 1"ϕ bolt

$$f_v = \frac{1.95}{4(0.785)} = 0.62 \text{ ksi}$$

$F_t = 59 - 1.9(f_v) = 59 - 1.9(0.62) = 57.8 > 45.0$ [Table J3.5]

use $F_t = 45$

$\phi F_t = 0.75 \times 45 = 33.8$ ksi

$$f_t = \frac{23.4(12)}{11.3(0.785)} = 31.7 \text{ ksi} < 33.8$$ Use 1"ϕ bolt

THE CORRECT ANSWER IS: (B)

512. $e = \frac{m}{p} = \frac{75}{55} = 1.36 \text{ ft} \quad 6(1.36) = 8.2 \text{ ft}$

Try 8.2 ft with resultant at kern $\frac{P}{A} + \frac{M}{S} = \frac{55}{67.2} + \frac{75(6)}{(8.2)^3} = 1.6$, so resultant is outside kern

$x = 3\left(\frac{L}{2} - e\right)$ and $q = \frac{2P}{Bx}$

Try 6 ft, x = 4.92, B = 6 ft, so q = 3.72 ksf

Try 5.5 ft, x = 4.17, B = 5.5 ft, so q = 4.79 ksf

Interpolate

Answer = 5.9 ft

THE CORRECT ANSWER IS: (C)

513. $F_{CE} = K_{CE} \frac{E'}{\left(L_e/d\right)^2}$

$L_e = K_e L = (1.0)(30)(12) = 360 \text{ inches}$

$L_e/d = \frac{360}{10.75} = 33.5 < 50 \quad \text{OK } (3.7.1.4)$

K_{CE} for glulam = 0.418

$E' = (C_m)(C_t)(E_{yy})$ (Tables 2.3.1, 2.3.4; Set 5.1.5)

$= (1.0)(1.0)(1.4)(10^6) = (1.4)(10^6) \text{ psi}$

$F_{CE} = 0.418 \frac{(1.4)(10^6)}{(33.5)^2} = 521 \text{ psi}$

THE CORRECT ANSWER IS: (A)

514. Check uplift, $0.85D = (0.85)(400) = 340$ kips > W or E uplift O.K.
Moment is small compared to axial load.

Maximum axial load, $P_{D+L+E} = (0.75)(400 + 150 + 250) = 600$ kips

Estimate footing size, $B = \sqrt{600/5} = 11$ ft

Check with 11'-6" footing size, $A = 11.5^2 = 132.25$ ft^2, $S = (11.5)^3/6 = 253.5$ ft^3

Case I, $P_{D+L} = 400 + 150 = 550$ kips

$M_{D+L} = 50 + 20 = 70$ ft-kips

$q = 550/132 + 70/253.5 = 4.4$ ksf < 5 ksf O.K.

Case II, $P_{D+L+W} = 0.75(400 + 150 + 150) = 525$ kips

$M_{D+L+W} = 0.75(50 + 20 + 150) = 165$ ft-kips

$q = 525/132 + 165/253.5 = 4.6$ kips < 5 ksf O.K.

Case III, $P_{D+L+E} = 600$ kips

$M_{D+L+E} = 0.75(50 + 20 + 0) = 53$ ft-kips

$q = 600/132 + 53/253.5 = 4.75$ ft-kips < 5 ksf O.K.

Use 11'-6"

THE CORRECT ANSWER IS: (A)

515. Weight of footing = $(1{,}560)(8)^2(3) = 29$ kips

$P_{D+L} = 23 + 29 = 52$ kips

$e = 150/52 = 2.9 \text{ ft} > B/6 = 8/6 = 1.33$ ft

Use only effective base,

$B_{eff} = 3(B/2 - e) = 3(8/2 - 2.9) = 3.3$ ft

$q_{max} = 2P/B_{eff} = 2(52)/(8 \times 3.3) = 3.9$ ksf

THE CORRECT ANSWER IS: (B)

516. Resistance by end bearing, $P_e = (30)(3.14/4)(14/12)^2 = 32$ kips/pile

Resistance by friction, $P_f = (1.5)(3.14)(14/12) = 5.5$ kips/ft

Length of friction = $[1{,}000 - (32 \times 4)]/(5.5 \times 4) = 39.6$ ft

Length of pile = $39.6 + 14 - 1 - 3 + 4/12 = 50$ ft

THE CORRECT ANSWER IS: (C)

517.

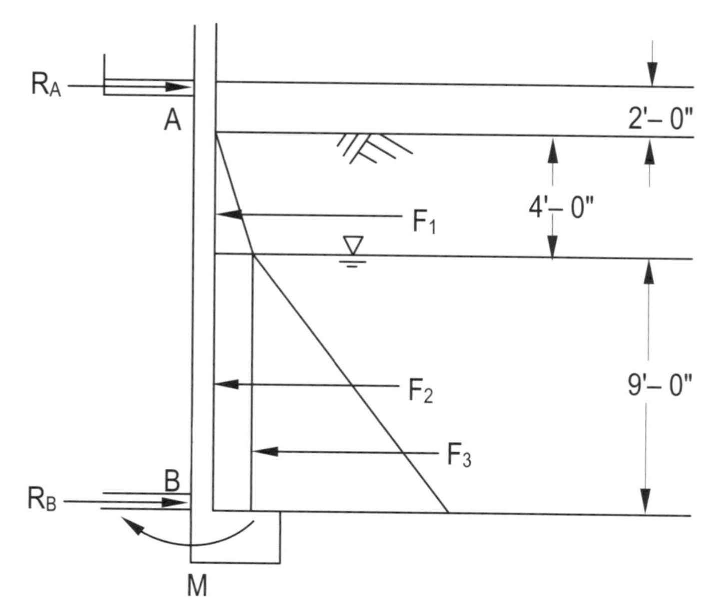

Use at rest pressure,

$F_1 = (1/2)\,(45)\,(4)^2 = 360$ lb/ft

$F_2 = (45)\,(4)\,(9) = 1{,}620$ lb/ft

$F_3 = (1/2)\,(35 + 62.4)\,(9)^2 = 3{,}945$ lb/ft

$\Sigma F = F_1 + F_2 + F_3 = 360 + 1{,}620 + 3{,}945 = 5{,}925$ lb/ft

THE CORRECT ANSWER IS: (D)

518. Use active pressures, as follows:

For surcharge $F_1 = (27)\,(2)\,(15) = 810$ lb/ft

For backfill $F_2 = (1/2)\,(27)\,(15)^2 = 3{,}038$ lb/ft

Total lateral force $= F_1 + F_2 = 810 + 3{,}038 = 3{,}848$ lb/ft

THE CORRECT ANSWER IS: (B)

519. Determine the minimum number of truckloads from the borrow pit.

The embankment construction requires 500,000 yd^3 of soil at:

$$\gamma_{dry} = (0.90)\,(120.0) = 108.0 \text{ pcf}$$

The total weight of dry soil required is:

$$W_{total} = (500{,}000 \text{ yd}^3)\,(27 \text{ ft}^3/\text{yd}^3)\,(108.0 \text{ pcf}) = 1.458 \times 10^9 \text{ lb}$$

The dry unit weight of soil in the truck is:

$$\gamma_{dry} = G_s\,\gamma_w/(1 + e) = (2.65)\,(62.4)\,/(1 + 1.30) = 71.9 \text{ psf}$$

Each truck can carry a weight of:

$$W_{truck} = (5.0 \text{ yd}^3)\,(27 \text{ ft}^3/\text{yd}^3)\,(71.9 \text{ pcf}) = 9{,}700 \text{ lb/truck}$$

Therefore, the minimum number of trucks required is:

$$N = W_{total}/W_{truck} = 1.458 \times 10^9/9{,}700 = 150{,}000 \text{ trucks}$$

THE CORRECT ANSWER IS: (B)

520. B/C Plan II to present

Benefit = reduction annual costs

= 250,000 – 248,000 = $2,000

$$\text{Costs} = 9{,}000\left(\frac{4}{p}\right)_{5\text{ yr}}^{10\%} + 1{,}000\,(0.10) \text{ or } 10{,}000\left(\frac{4}{p}\right)_{5\text{ yr}}^{10\%} - 1{,}000\left(\frac{4}{p}\right)_{5\text{ yr}}^{10\%}$$

$$= 2{,}374 + 100 = 2{,}474 \qquad\qquad 2{,}638 - 164 = 2{,}474$$

$$B/C = \frac{2{,}000}{2{,}474} = 0.81$$

THE CORRECT ANSWER IS: (B)

TRANSPORTATION

AFTERNOON SAMPLE SOLUTIONS

CORRECT ANSWERS TO THE TRANSPORTATION AFTERNOON SAMPLE QUESTIONS

Detailed solutions for each question begin on the next page.

501	A
502	C
503	D
504	B
505	B
506	D
507	D
508	B
509	A
510	C
511	B
512	D
513	B
514	A
515	A
516	A
517	D
518	C
519	C
520	B

TRANSPORTATION AFTERNOON SAMPLE SOLUTIONS

501. Referring to Chapter 9, HCM (1997 update) $G_p = 7.0 + \left(\frac{W}{4.0}\right) - Y$

W = distance from the curb to the center of the farthest travel lane
Y = 5.0 seconds
$W_{E\text{-}W} = 60$ ft
$W_{N\text{-}S} = 77.5$ ft

$$G_{pN\text{-}S} = 7.0 + \left(\frac{77.5}{4.0}\right) - Y$$

7.0 + 19.4 – 5.0 = 21.4 or most nearly 22.0 seconds

THE CORRECT ANSWER IS: (A)

502. Compute the "jam density."
The jam density is defined as the density at zero speed. Under these conditions, the vehicles are spaced at close intervals in the urban arterial street lane. Assuming the average spacing between the front bumpers of successive vehicles is approximately 30 feet, the jam density can be estimated as follows:

Jam density	(5,280 feet per mile)/(30 feet per vehicle)
Jam density	176 vehicles per mile
Rounding jam density	170 vehicles per mile

THE CORRECT ANSWER IS: (C)

503. Compute annual number of passenger cars.

Compute the total number of vehicles.

Let X = total number of vehicles.

(0.8500)(X)(2) = total number of passenger cars (2 axles)

(0.10)(X)(3) = total number of 3-axle trucks

(0.03)(X)(4) = total number of 4-axle trucks

(0.02)(X)(5) = total number of 5-axle trucks

24,560,000 axles = (0.8500)(X)(2) + (0.10)(X)(3) + (0.03)(X)(4) + (0.02)(X)(5)

24,560,000 axles = 2.22 X

X = 11,063,063 total vehicles

Compute Average Annual Daily Traffic (AADT)

AADT = total number of vehicles per year/days per year

AADT = 11,063,063/365 = 30,310 vehicles

THE CORRECT ANSWER IS: (D)

504. Determine the stopping sight distance.

From AASHTO Green Book, 1990, p. 289

For S<L

$$L = \frac{AS^2}{400 + 3.5(S)}$$

where S = stopping sight distance

A = algebraic difference in grades

L = length of curve

$$900 = \frac{8(S)^2}{400 + 3.5(S)}$$

Solve for S

S = 486.285 feet

THE CORRECT ANSWER IS: (B)

505. Determine the present worth of Alternate B.

Calculate the present worth for the annual maintenance.

P_{AM} = (annual maintenance for first 10 yr) (P/A, 10%, 10 yrs)

+ (annual maintenance for second 10 yr) ((P/A, 10%, 10yrs) (P/F, 10%, 10 yrs)

= $50,000 (6.1446) + $75,000 (6.1446) (0.3855) = $307,000 + $178,000 = $485,000

Calculate the present worth for the major maintenance.

P_{MM} = (major maintenance) (P/F, 10%, 10 yrs)

= $300,000 (0.3855) = $116,000

Calculate the present worth of the residual value.

P_{RV} = (residual value) (P/F, 10%, 20 yrs)

= $3,000,000 (0.1486) = $446,000

Calculate the sum, noting that the present worth of the first cost P_{FC} = $6,000,000 and that the residual value is a deduct.

Present worth = $P_{FC} + P_{AM} + P_{MM} - P_{RV}$

= $6,000,000 + 485,000 + 116,000 – 446,000

= $6,155,000 say $6,150,000

THE CORRECT ANSWER IS: (B)

506. Determine the traffic volume in the year 2000.

Use the interest tables with an interest rate of 5% for 6 years; i.e., from 1994 to 2000.

V_{2000} = V_{1994} (F/P, 5%, 6 yrs) = 30,000 veh/day (1.340) = 40,000 veh/day

THE CORRECT ANSWER IS: (D)

507. Pounds of cement per yd^3

$$6.28\ \text{sacks}/\text{yd}^3 \times 94\ \text{lb}/\text{sack} = 590.3\ \text{lb}/\text{yd}^3$$

Volume water in 1 yd^3 of concrete

1 yd^3 = water + cement + fine aggregate + coarse aggregate

$$\text{Absolute volume of cement/yd}^3 = 590.3\ \text{lb}/\text{yd}^3 \times \frac{1}{3.15} \times \frac{1\ \text{ft}^3}{62.4\ \text{lb}}$$

$$= 3.00\ \text{ft}^3/\text{yd}^3$$

$$\text{Absolute volume fine aggregate} = \frac{2.25 \times 590.3\ \text{lb}}{2.65} \times \frac{1}{62.4\ \text{lb}} = 8.03\ \text{ft}^3$$

$$\text{Absolute volume coarse aggregate} = \frac{3.25 \times 590.3\ \text{lb}}{2.65} \times \frac{1}{62.4\ \text{lb}} = 11.60\ \text{ft}^3$$

$$\text{Volume water} = 27.0 - 3.0 - 8.03 - 11.60$$

$$= 4.37\ \text{ft}^3 \times 7.48\ \text{gal/ft}^3 = 32.69\ \text{gal/yd}^3\ \text{concrete}$$

$$w/c = \frac{32.69\ \text{gal}}{6.28\ \text{sack}} = 5.2\ \text{gal/sack}$$

THE CORRECT ANSWER IS: (D)

508. B/C Plan II to present

Benefit = reduction annual costs

= 250,000 – 248,000 = $2,000

$$\text{Costs} = 9{,}000\left(\frac{4}{p}\right)^{10\%}_{5\ \text{yr}} + 1{,}000\,(0.10) \text{ or } 10{,}000\left(\frac{4}{p}\right)^{10\%}_{5\ \text{yr}} - 1{,}000\left(\frac{4}{p}\right)^{10\%}_{5\ \text{yr}}$$

$$= 2{,}374 + 100 = 2{,}474 \qquad 2{,}638 - 164 = 2{,}474$$

$$B/C = \frac{2{,}000}{2{,}474} = 0.81$$

THE CORRECT ANSWER IS: (B)

509. Determine the rate of change in degree of curvature, D (degrees per station), along the spiral. Compute rate of change of degree of curvature per foot of spiral (K).

$K = Dc/Ls$ (Hickerson, *Route Location and Design*, p. 169)

$= 3°/250\text{ ft} = 0.012°\text{ ft}$

Convert rate of change per foot to rate of change per station (Ks).

$Ks = (100\text{ ft/sta}) \times K$

$= (100\text{ ft/sta}) \times 0.012° = 1.20$

THE CORRECT ANSWER IS: (A)

510. Determine the station at the high point on Curve 1.

Compute the rate of change of grade, r, for Curve 1.

$r = (g_2 - g_1) / L = [-3.0\% - (+4.5\%)] / 14\text{ sta.} = -0.5357\% / \text{sta.}$

Compute the distance from the P.V.C.$_1$ to the high point on Curve 1.

$X_{PVC} = -g_1 / r = -(+4.5\%) / (-0.5357\%/\text{sta}) = 8.4002$ stations

Compute the P.V.C$_1$ station on Curve 1.

$P.V.C._1 = P.V.I._1 - L/2 = (42+00) - (7+00) = 35+00.00$

Compute the station of the high point on Curve 1.

High point station $= P.V.C._1 + X_{PVC}$

$= (35+00) + (8+40.02)$

$= 43+40.02$ which is most nearly 43+40

THE CORRECT ANSWER IS: (C)

511. Compute the tangent elevation at station 73+00.

Tan. elev. = 334.56 feet + (3.0)(3.0) = 343.56 feet

Compute the tangent offset, e, at the P.V.I. station on Curve 2.

$e = (L/8)(g_2 - g_1) = (15/8)\,[+2.0\% - (-3.0\%)] = 9.375$ feet

Compute the tangent offset, y, at station 73+00.

$y = (4e / L^2)\,X^2 = [(4)\,(9.375)\,(4.5)^2] / (15)\,(15) = 3.38$ feet

Compute the vertical curve elevation at station 73+00.

Curve elevation = Tangent elevation + tangent offset.

= 343.56 + 3.38 = 346.94 feet

Compute the clearance between the bridge and the vertical curve at station 73+00 as follows:

Clearance = Bridge elevation – Curve elevation

= 365.94 – 346.94 = 19.0 feet

THE CORRECT ANSWER IS: (B)

512. The station where full superelevation is reached.

Compute the S.C. station, this is where the superelevation will occur.

S.C. Sta. = T.S. Sta. + Ls

= (50 + 00) + (2 + 50) = 52 + 50

THE CORRECT ANSWER IS: (D)

513. SSD in feet = P – R distance plus braking distance

$$SSD = 2.5\,(1.47)\,(45) + \frac{(45)^2}{30\,(0.38 - 0.01)}$$

SSD = 165 ft + 182 ft = 347 ft

THE CORRECT ANSWER IS: (B)

514. Option (A) is correct; it is **NOT** true. On nuclear gauges requiring a badge, the operator must also be certified. A safety course must be completed.

THE CORRECT ANSWER IS: (A)

515. Option (A) is correct; this statement is **NOT** true. If the water-to-cement ratio is decreased, the amount of excess water is lesser; thus the lesser is the amount of pores and interconnected capillaries in the concrete. Lower porosity improves water-tightness; water-tightness is increased. See the PCA Design & Control of Concrete mixtures, p. 7–8, *Materials Science in Engineering*, Keyser, p. 322, or other sources.

THE CORRECT ANSWER IS: (A)

516. Option (A) is correct; it is **NOT** true that the hotter the mix above 300°F, the better the results. If a mix is too hot it will push up in front of the roller and cause poor compaction. The mix is usually between 260 and 280°F. (*Asphalt Handbook*, p. 289)

THE CORRECT ANSWER IS: (A)

517. Determine the depth of water in the trapezoidal ditch.
For a trapezoidal channel with a depth = y, base = b = 2.0 ft, and side slopes 3:1 (horizontal:vertical),
$z = 3$:

Area $= A = (b + zy)y = (2.0 + 3y)y$

$$R = \frac{(b + zy)y}{b + 2y\sqrt{1 + z^2}} = \frac{(2.0 + 3y)y}{2.0 + 2y\sqrt{10}}$$

With n = 0.02, s = 0.005, and Q = 30 cfs, use the trial and error method to solve the Manning Equation for the depth, y:

$$Q = \frac{1.486}{n} A\, R^{2/3} s^{1/2} = \frac{1.486}{0.02} A\, R^{2/3} (0.005)^{1/2} = 30.0$$

Trial	Depth (inches)	Depth (feet)	R (feet)	A (ft^2)	Q (cfs)
1	12	1.00	0.601	5.00	18.7
2	15	1.25	0.726	7.188	30.5

The depth is most nearly 15 inches.

THE CORRECT ANSWER IS: (D)

518. Determine the rectangular channel's flow velocity.
Flow area = 3.9 × 1.9 = 7.4 ft^2
V = Q/A = 39/7.4 = 5.3 fps

THE CORRECT ANSWER IS: (C)

519. For a 10-year storm and for T_c of 7.5 minutes:

$I = (2.95 + 2.08)/2 = 2.5$ in/hr

From **Figure 2**, C = 0.82

THE CORRECT ANSWER IS: (C)

520. Determine the peak runoff flow for a 25-year storm at Point 2. At Point 2 in the watershed, the corresponding T_c is controlled by the maximum T in the drainage basin plus the travel time in the drainage ditch. The maximum T_c for the individual drainage basins associated with Basin B is 35 minutes.

$$\text{Travel Time} = \frac{1{,}900 \text{ ft}}{(1.7 \text{ ft/s})(60 \text{ s/min})} = 19 \text{ min}$$

$T_c = 35 + 19 = 54$ min

The corresponding intensity for a 25-year event is determined from **Figure 2** to be 0.77 in/hr. The runoff coefficient is determined as an area-weighted average for the entire basin upstream of Point 2:

$$C = \frac{\sum A\,C}{\sum A} = \frac{(8)(0.1)+(12)(0.1)+(6)(0.15)+(6)(0.4)+(4)(0.4)+(4)(0.6)}{(8+12+6+6+4+4)} = 0.232$$

thus

$$Q = C\,i\,A$$

$$A = 8 + 12 + 6 + 6 + 4 + 4 = 40 \text{ acres}$$

$$Q = (0.2325)(0.77 \text{ in/hr})(40 \text{ acres})\left(\frac{1.008 \text{ cfs}}{\text{acre} - \text{in/hr}}\right) = 7.22 \text{ cfs}$$

Similar calculations are noted for determination of peak flows at Point 2 originating from individual drainage basins.

Basin	Travel Time (min)	T_c (min)	i (in/hr)	C	A (acres)	Q (cfs)
F	4	14	1.70	0.60	4	4.1
EF	4	19	1.45	0.50	8	5.8
DEF	9	29	1.12	0.46	14	7.2
CDEF	9	34	1.01	0.36	20	7.4
ACDEF	19	49	0.80	0.29	28	6.5
ABCDEF	19	54	0.77	0.23	40	7.2

The maximum flow is selected as the likely peak flow, corresponding with the drainage area CDEF, Q = 7.4 cfs which is most nearly 7.5 cfs.

THE CORRECT ANSWER IS: (B)

WATER RESOURCES
AFTERNOON SAMPLE SOLUTIONS

CORRECT ANSWERS TO THE WATER RESOURCES AFTERNOON SAMPLE QUESTIONS

Detailed solutions for each question begin on the next page.

501	D
502	A
503	D
504	A
505	B
506	A
507	D
508	A
509	A
510	B
511	C
512	C
513	B
514	A
515	A
516	C
517	C
518	A
519	C
520	D

WATER RESOURCES AFTERNOON SAMPLE SOLUTIONS

501. For the pipe network system, the pressure at any point in the system depends on all the factors that contribute to the total energy: flow conditions, energy losses, and elevations.

Option (A) is incorrect because the pressure at Point D will be influenced by the energy loss in pipe segment AD.

Option (B) is incorrect because the pressure depends on balanced flows that are not supplied. If the head loss in pipe AD is small relative to the difference in elevation (Point D is 20 feet below Point A), the pressure at Point D would be greater than the pressure at Point A.

Option (C) is incorrect because elevations are not specified for Points B and C. Therefore the pressures are indeterminate (although the sum of elevation and pressure head could be determined.)

Option (D) is correct. Development of an energy equation between Points A and D can be used to determine the pressure at Point D, provided that the stated items are known: total energy at Point A, friction loss between Points A and D, and elevation at Point D.

THE CORRECT ANSWER IS: (D)

502. Water hammer analysis per Metcalf and Eddy, *Wastewater Engineering Collection and Pumping of Wastewater*, McGraw Hill, 1981, pp. 389–398.

Assume the water main is buried. Specified conditions include:

L (pipe length)	1,000 feet
K (bulk modulus for water)	300,000 psi
d (pipe diameter)	8 inches
E (modulus of elasticity for pipe)	24,000,000 psi
e (pipe wall thickness)	0.27 inches

The critical period for the valve closure is calculated by:

$$T = \frac{2L}{a}$$

in which

T = critical period for valve closure
a = velocity of pressure wave

The pressure wave velocity is calculated by:

$$a = \frac{4{,}720 \text{ fps}}{\left[1 + C\frac{K}{E}\frac{d}{e}\right]^{1/2}}$$

in which

$C = 1 - \mu^2$ for buried force main (Metcalf and Eddy, 1981, p. 390)
μ = Poisson's ratio = 0.3 (Metcalf and Eddy, 1981, p. 391)

thus

$$a = \frac{4{,}720 \text{ fps}}{\left[1 + \left(1 - (0.3)^2\right)\frac{(300{,}000)(8)}{(24{,}000{,}000)(0.27)}\right]^{1/2}}$$
$$= 4{,}082 \text{ fps}$$

thus

$$T = \frac{2L}{a} = \frac{2\ (1{,}000 \text{ ft})}{4{,}082 \text{ fps}} = 0.49$$

THE CORRECT ANSWER IS: (A)

503. The following information is specified:

Q_{peak} = 6 MGD = 9.284 cfs
Q_{min} = 0.70 MGD = 1.083 cfs
n = 0.011
s = 0.001

The Manning equation is used to calculate flow:

$$Q = \frac{1.486}{n} A R^{2/3} s^{1/2}$$

in which

Q = discharge (cfs)
n = Manning roughness coefficient
A = flow area (ft2)
R = hydraulic radius (ft)
s = channel slope

For a circular pipe flowing full:

$$A = \frac{3.1416 D^2}{4}$$

$$R = \frac{D}{4}$$

in which

D = pipe diameter (ft)

$$Q = \frac{1.486}{n} \frac{3.1416 D^2}{4} \left(\frac{D}{4}\right)^{2/3} s^{1/2} = \frac{0.463}{n} D^{8/3} d^{1/2}$$

$$D = \left(\frac{Qn}{0.463 s^{1/2}}\right)^{3/8}$$

For the critical conditions to prevent surcharges (Q_{max}):

$$D = \left[\frac{(9.284)(0.011)}{(0.463)(0.001)^{1/2}}\right]^{3/8}$$

$$= 2.07 \text{ ft}$$

$$= 24.9 \text{ inches}$$

A larger pipe must be selected to avoid surcharged conditions. The most economical pipe is the next largest available size.

THE CORRECT ANSWER IS: (D)

504. For the problem specifications, n is assumed to be constant. For the minimum flow conditions, Q_{min} = 1.083 cfs. The flowing full capacity of the 36-inch sewer is 24.9 cfs. Corresponding flow equations are presented for the two conditions:

$$Q_{min} = 1.083 = \frac{1.486}{n} A\, R^{2/3} s^{1/2}$$

$$Q_{full} = 24.9 = \frac{1.486}{n} \frac{3.1416\, D^2}{4} \left(\frac{D}{4}\right)^{2/3} s^{1/2}$$

Thus, taking the ratio of these two equations:

$$\frac{Q_{min}}{Q_{full}} = \frac{1.083}{24.9} = 0.0435 \frac{A\, R^{2/3}}{0.3117\, D^{8/3}}$$

$$\frac{A\, R^{2/3}}{D^{8/3}} = (0.0435)(0.3117)$$

$$= 0.01355$$

From geometric elements for circular channel sections (Chow, 1959, p. 625), this flow condition corresponds with a depth of the flow equal to 14% of the pipe diameter. Thus:

depth = 0.14
D = 0.14 (36 inches) = 5.04 inches

The answer can be obtained with graphical procedures (n = constant with depth using a hydraulic element chart (WPCP, 1969, p. 89)

$$\text{If } \frac{Q}{Q_{full}} = 0.0435$$

$$\text{then } \frac{\text{depth}}{D} = 0.14$$

THE CORRECT ANSWER IS: (A)

505. Determine the flow rate if there were two pumps in parallel in the existing 10-inch pipe system. The pump curve for two pumps operating in parallel is obtained from the pump curve for the single pump by doubling the flow for each value of TDH in **Figure 2** (the pump characteristics curves). The system curve is calculated as shown below:

TDH (ft)	One Pump Q (gpm)	Two Parallel Pumps 2 Q (gpm)
68	200	400
66	400	800
62	600	1,200

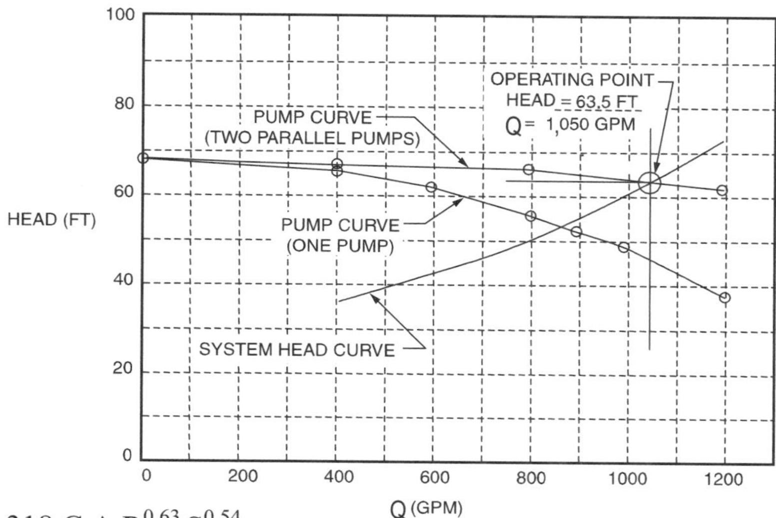

$Q = V A = 1.318\ C\ A\ R^{0.63}\ S^{0.54}$
$C = 100$
$A = (1/4)\ \pi D^2 = (1/4)(3.1416)(0.833)^2 = 0.545\ ft^2$
$R = D/4 = (0.833/4) = 0.208\ ft$

$$S^{0.54} = \frac{Q}{1.318\ C\ A\ R^{0.63}} = \frac{Q}{1.318(100)(0.545)(0.208)^{0.63}} = 0.03744\ Q$$

$S = 0.00228\ Q^{1.8519}$
$h_f = (3{,}000\ ft)(0.00228)\ Q^{1.8519}$
$TDH = \Delta H + h_f + V^2/2g$ and $\Delta H = 560\ ft - 530\ ft = 30\ ft$

Q (gpm)	Q (cfs)	h_f (ft)	V = Q/A (fps)	$V^2/2g$ (ft)	ΔH (ft)	TDH (ft)
400	0.891	5.524	1.635	0.042	30.00	35.6
600	1.337	11.713	2.453	0.093	30.00	41.8
800	1.783	19.962	3.272	0.166	30.00	50.1
1,000	2.228	30.158	4.088	0.260	30.00	60.4
1,200	2.674	42.282	4.906	0.373	30.00	72.7

Plot the new pump curve for two pumps in parallel plus the system curve. The operating point is located where the new pump curve crosses the 10-inch system head curve.

THE CORRECT ANSWER IS: (B)

WATER RESOURCES AFTERNOON SAMPLE SOLUTIONS

506. Determine the maximum elevation of the pump to avoid cavitation when operating at maximum efficiency. From **Figure 2**, at maximum pump efficiency = 82%, the TDH = 52.0 ft, and NPSH = 7 ft. Prorate the total head loss for the suction side.

Total h_f = TDH – ΔH = 52.0 – (560 – 530) = 22 ft
Suction side h_f = 22 ft (30 ft/3,000 ft) = 0.2 ft

At a pump elevation of 530 ft, $(P/\gamma) = (P_{atm}/\gamma - P_{vapor}/\gamma)$ = (33.3 ft – 0.6 ft) = 32.7 ft of water.

$$NPSH = (P/\gamma) + \Delta Z - h_f$$
$$= 32.7 + (530 - EL_{pump}) - 0.2 = 7$$
$$EL_{pump} = 530 + 32.7 - 0.2 - 7 = 555.5 \text{ ft}$$

THE CORRECT ANSWER IS: (A)

507. Determine which statement is true regarding valves.

I. Flow control valves should not be located on the suction side of the pump.
III. A check valve should be installed on the suction side of the pump to maintain pump prime when inactive.

THE CORRECT ANSWER IS: (D)

508. Determine the average depth of water that will run off. Appropriate equations are from Gupta, 1989, pp. 100–103:

Q = runoff (inches)
CN = curve number
S = potential maximum retention of water by soil (inches)
P = accumulated rainfall (inches)

For CN = 50

$$CN = \frac{1{,}000}{10 + S} = 50$$

Solving for S,

$$S = \frac{1{,}000 - (10)(50)}{50} = 10 \text{ inches}$$

For S = 10 inches and P = 5.0 inches, the runoff depth is:

$$Q = \frac{(P - 0.2\,S)^2}{(P + 0.8\,S)} = \frac{(5.0 - 0.2 \times 10)^2}{(5.0 + 0.8 \times 10)} \qquad Q = 0.69 \text{ inch}$$

THE CORRECT ANSWER IS: (A)

509. Determine the peak runoff flow for a 25-year storm at Point 1. For Point 1 in the watershed, the time to concentration (T_c) is controlled by the maximum T_c in the drainage basin plus the travel time in the drainage ditch. The maximum T_c for the individual drainage basins is associated with Basin B (35 minutes as specified in the table.) The travel time in the drainage ditch is calculated from the length of the ditch (1,000 ft) and the velocity (1.7 fps).

$$\text{Travel time} = \frac{1{,}000 \text{ ft}}{(1.7 \text{ fps})(60 \text{ sec/min})}$$

$$= 9.8 \text{ min}$$

$$T_c = 35 + 9.8$$

$$= 44.8 \approx 45 \text{ min}$$

For the specified design storm (25-year frequency) and the calculated T_c (45 min), the rainfall intensity is determined from **Figure 2** to be 0.85 in./hr. The runoff is calculated from the Rational Formula:

$$Q = C\,i\,A$$

For the problem specifications, C = 0.1 for the drainage area upstream from Point 1 (Basins A and B) i = 0.85 in./hr, and A = 8 + 12 = 20 acres.

$$Q = (0.1)(0.85 \text{ in./hr})(20 \text{ acres})\left(\frac{1.008 \text{ cfs}}{\text{acre-in./hr}}\right)$$

$$Q = 1.7 \text{ cfs}$$

THE CORRECT ANSWER IS: (A)

510. Determine the unit hydrograph discharge at Hour 10. The procedure for determination of the unit hydrograph follows the method of Gupta (1989, pp. 302–304 and 310–311). The flow data indicate zero discharge prior to the rainfall event. The flow returns to zero after 14 hours. Based on these zero flows, it may be concluded that the base flow is zero. The total runoff volume (V) is obtained by integration of the flow vs. time response. The unit hydrograph for an event of equal duration is obtained by division of the measured flow values by the runoff depth, defined as the runoff volume divided by the drainage area.

Time (hr)	Runoff (cfs)	Avg. Runoff (cfs)	Duration (hr)	Volume (cfs-hr)
0	0.0	--	--	--
2	2.5	1.25	2	2.50
4	5.0	3.75	2	7.50
6	7.5	6.25	2	12.50
8	4.0	5.75	2	11.50
10	2.0	3.00	2	6.00
12	1.0	1.50	2	3.00
14	0.0	0.50	2	1.00
Sum	0.0		2	44.00
				44.00

$$\text{Runoff Depth} = \frac{(44 \text{ cfs-hr})(3{,}600 \text{ sec/hr})}{(40 \text{ acres})(43{,}560 \text{ ft}^2/\text{acre})} = 0.091 \text{ ft} = 1.1 \text{ inches}$$

The measured discharge at Hour 10 = 2 cfs. For a unit hydrograph (1-inch event), the corresponding discharge (Q) is calculated as:

Q = (2 cfs) (1 inch)/(1.1 inches) = 1.8 cfs

THE CORRECT ANSWER IS: (B)

511. Determine the runoff coefficient for the entire basin. The average runoff coefficient is determined with an area-weighted average of the runoff coefficient for each parcel in the drainage basin:

$$C = \frac{\sum A\,C}{\sum A} = \frac{(8)(0.1)+(12)(0.1)+(6)(0.15)+(6)(0.4)+(4)(0.4)+(4)(0.6)}{(8+12+6+6+4+4)}$$

$$C = 0.23$$

THE CORRECT ANSWER IS: (C)

512. Hazen-Williams equation

$V = 1.318\, C\, R^{0.63}\, S^{0.54}$

$Q = VA$

$A = Q/A$

$Q = 1.318\, CAR^{0.63}\, S^{0.54}$

$C = 100$

$$A_{10} = \pi\frac{D^2}{4} = 3.1416\left(\frac{0.83^2}{4}\right) = 0.5411 \text{ ft}$$

$$R_{10} = \frac{D}{4} = \frac{0.83}{4} = 0.2083 \text{ ft}$$

$$S^{0.54} = \frac{Q}{1.318\, CAR^{0.63}}$$

$$S^{0.54}\ \frac{Q}{1.318(100)(0.5411)(0.2083)^{0.63}}$$

$$S_{10}^{0.54} = 0.0377\, Q$$

$$S_{10} = 0.0023\, Q^{1.8519}$$

$$h_{f_{10}} = 3{,}000(0.0023)(Q^{1.8519})$$

$TDH = \Delta H + h_f + v^2/2g$

10-inch pipe

Q (gpm)	Q (cfs)	h_f (ft)	v (fps)	$v^2/2g$ (ft)	ΔH (ft)	TDH
0	0	0	0	0	30	30
200	0.45	1.51	0.81	0.01	30	31.5
400	0.88	5.45	1.63	0.04	30	35.5
600	1.32	11.5	2.44	0.09	30	41.6
800	1.76	19.7	3.25	0.16	30	49.8
900	1.98	24.5	3.66	0.21	30	54.6
1,000	2.20	29.7	4.01	0.26	30	59.9

Plot system head curve (TDH). Where the system head curve crosses the pump curve is the point of operation.

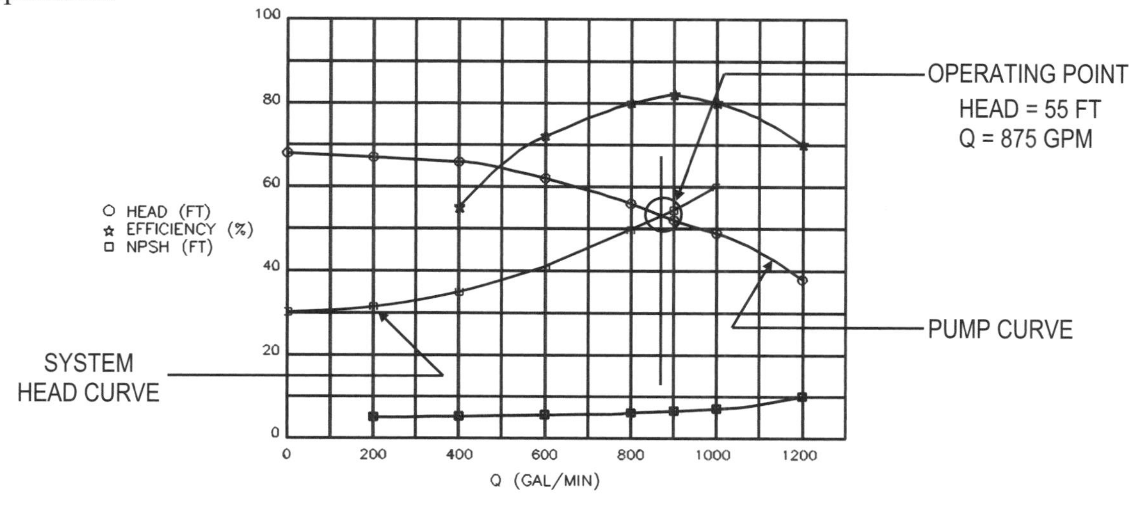

PUMP CHARACTERISTICS

Q = 875 gpm

THE CORRECT ANSWER IS: (C)

513. The treatment plant flow rate, Q = 6 MGD. All flow will go through the rapid-mix basins and the design detention time t_d = 30 sec. $t_d = V/Q$, V = basin volume. $V = t_d \times Q$.
$V = 30 \text{ sec} \times 6 \times 10^6 \text{ gpd} \times 1 \text{ day}/24 \text{ hr} \times 1 \text{ hr}/60 \text{ min} \times 1 \text{ min}/60 \text{ sec} \times 1 \text{ ft}^3/7.48 \text{ gal}$
$V = 278.5 \text{ ft}^3$

Maximum size of each unit is 50 ft^3.
Number of units = 278.5 ft^3/50 ft^3 = 5.6 units; use 6 units
Volume of rapid mixer = 278.5 ft^3/6 = 46.4 ft^3

THE CORRECT ANSWER IS: (B)

514. Determine the reduction of BOD_5 at the plant.

Wastewater flow rate = (30,000 persons) (100 gal/person/day) (MG/10^6 gal)
= (3.0 MGD) (1.5473 cfs/MGD)
= 4.64 cfs

Calculate the BOD_5 of the wastewater influent.

BOD_5 loading = (0.2 lb/person/day) (30,000 persons) = 6,000 lb/day

$$BOD_5 \text{ of the wastewater influent} = \frac{6{,}000 \text{ lb/day}}{(3.0 \text{ MGD})\left(\frac{8.34 \text{ lb/MG}}{\text{mg/L}}\right)} = 240 \text{ mg/L}$$

Do mass balance at discharge point

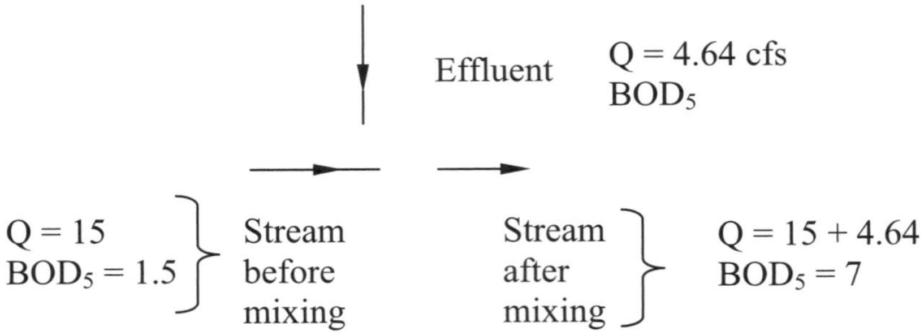

Effluent Q = 4.64 cfs, BOD_5

Q = 15, BOD_5 = 1.5 } Stream before mixing → Stream after mixing } Q = 15 + 4.64, BOD_5 = 7

$$\therefore BOD_5 \text{ of effluent} = \frac{[(15+4.64)\times 7 - 15\times 1.5]}{4.64}$$

$$= 24.8 \text{ mg/L}$$

$$\therefore \text{Required removal efficiency} = \left(\frac{240-24.8}{240}\right)100$$

$$= 89.7\%$$

THE CORRECT ANSWER IS: (A)

515. The required total chlorine residual is calculated with the following:

$$N_t = N_o (1 + 0.23\ C_t\ t)^{-3}$$

$$C_t = \frac{\left[\left(\frac{N_o}{N_t}\right)^{\frac{1}{3}} - 1\right]}{0.23\ t} = \frac{\left[\left(\frac{10{,}000}{200}\right)^{\frac{1}{3}} - 1\right]}{0.23(15)} = 0.778 \text{ mg/L}$$

THE CORRECT ANSWER IS: (A)

516. Assume the pKa for NH_3 is 9.3. The concentration (mg/L) of nonionized ammonia nitrogen (NH_3-N) 10 miles downstream from the point of release is most nearly:

$$\frac{[NH_3][H^+]}{[NH_4^+]} = 10^{-9.3}$$

$$\frac{[NH_3][10^{-7.5}]}{[NH_4^+]} = 10^{-9.3}$$

$$\frac{NH_3}{NH_4^+} = 10^{-1.8} = 0.0158$$

$$[NH_3] + [NH_4^+] = 10 \text{ mg/L}$$

$$0.0158[NH_4^+] + [NH_4^+] = 10 \text{ mg/L}$$

$$[NH_4^+] = 9.844$$

$$[NH_3] = (9.844)(0.0158) = 0.16 \text{ mg/L}$$

THE CORRECT ANSWER IS: (C)

517. Find the location of the critical DO deficit. The critical deficit is the point downstream where the DO value is the lowest.

Find $t_{critical}$ based on DO sag curve formula.

$$t_{critical} = \frac{1}{k_2 - k} \ln\left[\frac{k_2}{k}\left(1 - \frac{DO_{det}(k_2 - k)}{k\,L_o}\right)\right]$$

where $DO_{det} = DO_{sat}$ at 20° – DO_{ace} = 9.08 – 6 = 3.08 mg/L

$$t_c = \frac{1}{0.4 - 0.23} \ln\left[\frac{0.4}{0.23}\left(1 - \frac{3.08\,(0.4 - 0.23)}{0.23\,(10)}\right)\right] \frac{1}{\text{day}^{-1}}$$

t_c = 1.74 days

$t \times v$ = distance downstream

Distance = (1.74 days)(1 fps)(3,600 sec/hr)(24 hr/day)(1 mi/5,280 ft)

= 28.5 miles

THE CORRECT ANSWER IS: (C)

518. Given:
Confined aquifer at non-steady state conditions
T = 38,000 gpd/ft
S = 3×10^{-4}
T = 100 d
Well diameter = 16 inches ∴ radius = 8 inches

Theis equation:

$$s = \frac{114.6\ Q}{T}\int_u^\infty \frac{e^{-u}}{u}\,du = \frac{114.6\ Q}{T} \times W(u)$$

Where
s = Drawdown, ft

Q = Flow, gpm

T = Transmissivity, gpd/ft

$$u = \frac{1.87\ r^2\ S}{t\ T}$$

r = Well radius, ft

S = Storativity, unitless

W(u) = Well function of u

Calculate:

s = 1,420 ft – 1,400 ft = 20 ft

$$u = \frac{1.87\ r^2\ S}{t\ T} = \frac{1.87 \times (8/12)^2 \times 3 \times 10^{-4}}{100 \times 38{,}000} = 6.56 \times 10^{-11}$$

If $u < 0.01$, then $W(u) = -0.5772 - \ln u = -0.5772 - \ln(6.56 \times 10^{-11}) = 22.9$

$$s = \frac{114.6\ Q}{T} \times W(u)$$

$$20 = \frac{114.6\ Q}{38{,}000} \times 22.9$$

$$Q = 290\ \text{gpm}$$

THE CORRECT ANSWER IS: (A)

Reference:
Viessman and Hammer, *Water Supply and Pollution Control*, 6th ed., pp. 53–55

519. Constant Head Permeability Test:

V = volume = 10 in^3

L = sample length = 6 inches

A = area of sample = $1/4 \times \pi\,(3\text{ in})^2 = 7.07\text{ in}^2$

H = head loss = (30 inches – 10 inches) = 20 inches

t = time of flow = 2 minutes

$$k = \frac{VL}{AHt} = \frac{(10\text{ in}^3)\times(6\text{ in.})}{(7.07\text{ in}^2)\times(20\text{ in.})\times(2\text{ min})} = 0.21\text{ in./min}$$

THE CORRECT ANSWER IS: (C)

520. Falling Head Permeability Test:

a = cross-sectional area of burette = $1/4 \times \pi\,(1\text{ in.})^2 = 0.785\text{ in}^2$

L = length of sample = 6 inches

A = area of sample = $1/4 \times \pi\,(3\text{ in.})^2 = 7.07\text{ in}^2$

t = time of flow = 20 minutes

h_1 = initial head in burette = 20 inches

h_2 = final head in burette = 10 inches

$$k = \frac{aL}{At}\ln\left(\frac{h_1}{h_2}\right) = \frac{(0.785\text{ in}^2)\times(6\text{ in.})}{(7.07\text{ in}^2)\times(20\text{ min})}\times\ln\left(\frac{20\text{ in.}}{10\text{ in.}}\right) = 0.023\text{ in./min}$$

THE CORRECT ANSWER IS: (D)

APPENDIX A

SAMPLE OF EXAM COVERS AND INSTRUCTIONS

National Council of Examiners for Engineering and Surveying

Serial

NAME: ______________________________
Last First Middle Initial

CIVIL BREADTH SAMPLE

PRINCIPLES AND PRACTICE OF ENGINEERING EXAMINATION

Answers to multiple-choice questions must be placed on the separate answer sheet

SAMPLE
CIV–AM

Sample questions and solutions are also available for the following examinations:

Chemical Engineering
Electrical and Computer Engineering
Environmental Engineering
Mechanical Engineering
Structural I Engineering
Structural II Engineering

For more information about these and other Council study materials (including CD-ROMs and Internet Diagnostic tests), visit our homepage at www.ncees.org or contact our Customer Service Department at 800-250-3196.